T0371748

Ground Beetles and Their Role in Management of Crop Pests

Ground Beetles and Their Role in Management of Crop Pests

Chitta Ranjan Satpathi
Professor, Agricultural Entomology
Bidhan Chandra Krishi Viswavidyalaya
West Bengal, India

CRC Press is an imprint of the
Taylor & Francis Group, an **informa** business

Capital Publishing Company
NEW DELHI KOLKATA

First published 2022
by CRC Press
4 Park Square, Milton Park, Abingdon, Oxon, OX14 4RN

and by CRC Press
6000 Broken Sound Parkway NW, Suite 300, Boca Raton, FL 33487-2742

CRC Press is an imprint of Informa UK Limited

British Library Cataloguing-in-Publication Data
A catalogue record for this book is available from the British Library

Library of Congress Cataloging-in-Publication Data
A catalog record has been requested

ISBN: 9781032161204 (hbk)
ISBN: 9781003261087 (ebk)

DOI: 10.4324/9781003261087

Typeset in Times New Roman
by Innovative Processors, New Delhi - 110009

Preface

This book has been written keeping in mind to collect relevant information on the role of ground beetles on biological control of insect, mites, fungus, mollusk and weed pests of different crops, mass multiplication of some important predacious ground beetles, statistical analyses, preparation of artificial diets and their suitable application for their rearing in laboratories. The aim is to provide reasonable coverage of the subject as a whole. This is probably the first publication which may be used as hand book for identification and distribution of common species of predacious ground beetles around the world. It is hoped that this book would also be a guide book to researchers, students, teachers, extension workers and progressive farmers who are looking for information on the diverse topics of biological control. Scientific descriptions have been minimized so that the write up can be more easily understood. Colour photographs will provide easy identification of the species and thereby help to restrict the unnecessary use of chemical pesticides in crop fields. Excellent illustrations augmented with micrographs, and exhaustive references make it a valuable addition to all biologists' bookshelves.

In order to successfully tackle plant stress condition caused by the biotic factors, a plant protection practitioner must have knowledge of different agroecosystems with diverse interacting communities of the insect pests and their natural enemies population. I have consulted various textbooks, papers, theses, memoirs, and downloaded photographs from internet in the preparation of this book, to the authors of which, I am grateful. I am also grateful to my friends and professional colleagues specially Dr. Ashis Chakraborty, Professor in Agronomy, BCKV, Dr. Bijon Kumar Das, Professor in Agricultural Entomology, BCKV as well as my sons Sourin and Sandip for their help and advice and to all those whose comments and criticisms have helped to write this book with precise and up-to-date information and data base. Any mistakes or misinterpretations are those of mine, and I will happily receive comments and criticisms on aspects of the contents. The author will deem the hard labour and the time that went into the writing of the book amply rewarded if it proves beneficial to the readers.

Chitta Ranjan Satpathi
(csatpathi2003 @yahoo.co.in)

Contents

1

Introduction

1.1 Origin of Ground Beetle

Carabidae is one of the largest family in the order Coleoptera of class Insecta. The insects under this family are commonly known as ground beetles, comprising about 35,000 species (Lorenz 2005). The first fossil Coleoptera are known from the Permian (290 Ma ago) belonging to the suborder Archostemata. By the Late Triassic (200 Ma ago) Carabidae began to appear in the fossil record (Arnol' di et al. 1992). A fossil of a single fore-tibia from the Upper Cretaceous (140 Ma ago) exhibit a well-developed antennal cleaner characteristic of advanced Carabidae in the subfamily Harpalinae (sensu Crowson 1955) and by the early Cretaceous (140 Ma ago) a number of extant carabid subfamilies are recognised. Among the other predatory carabid the fossil record of Scaritinae was extremely poor and those that are known only date back as far as the middle Eocene (47 Ma ago) (Lutz 1990). *Scarites haldingeri* Heer was obtained from Lower Miocene rocks (Heer, 1861) and the genus *Glenopterus* was established by Heer 1847 for a scaritine fossil from Miocene deposits in Germany.

The insects are commonly found on the grounds, their elytra become fixed and hind wings are atrophied. It is also plenty in tropics where they are ground-dwellers and also arboreal. The hind-wing and elytra of this beetle become free. The members of this great family are mostly sombre in appearance and a few are brightly coloured. Majority of them are nocturnal and active during night-time. They usually live under bark, stone, logs, rotten wood especially on damp places. Their larvae and adults feed on other insects. Carabid beetles are an incredibly diverse group of insects with over 40,000 species worldwide, classified into some 86 tribes (Erwin 1985). A worldwide survey of the literature (Larochelle 1990) reporting on 1054 species of carabids and cicindelids show that 775 species (73.5%) are exclusively carnivorous, 85 species (8.1%) phytophagous, and 206 species (19.5%) omnivorous. In India, it is found in all places from top mountain area to sea-level. The relationships between the major lineages of Carabidae are summarized in Fig. 1.

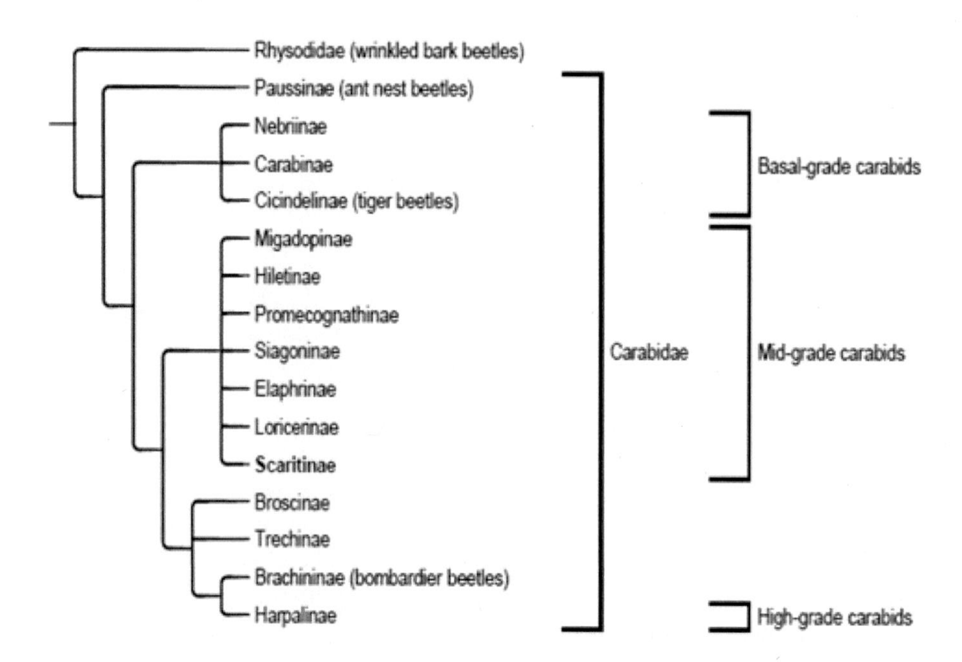

Fig. 1. Diagram illustrating relationships of the major lineages of 'grdaes' of Carabidae (after Erwin 1985 and Maddison et al. 1999).

Carabids emerged in the early tertiary as wet-biotope generalists in tropical habitats, where they are one of the dominant predatory invertebrate groups (Erwin & Adis 1982). Through a series of taxon pulses, they have radiated to drier environments as well as higher latitudes and altitudes (Erwin 1979a). By the late Permian-early Triassic, several lineages developed a cosmopolitan distribution pattern, as demonstrated by the fossil record (Ponomarenko 1977). Although this group have retained an easy to recognize generalist body plane, their body shape and leg morphology are characteristically modified for running, digging, burrowing, climbing and swimming (Evans 1977, 1986). Different parts of the morphological apparatus and physiological mechanisms can evolve different rates. Thus, a species can remain a generalist structurally and still became a specialist physiologically in order to, for example, live at glacier edges (*Nebria* spp.). Several other structural, physiological and behavioural adaptations enable carabids to invade all major habitats, where at least some lineages have attained dominance; the only exception is deserts, where carabids are limited to streams and oases (Erwin 1985). This distribution pattern suggests that humidity is a general limiting factor. The main structural patterns in carabid evolution are flightlessness and arboreal, fossorial, and troglobitic adaptations (Erwin 1985). Flightlessness have repeatedly evolved in many groups (Darlington 1943). In the tropics, >30% of species are arboreal, exhibit special morphological (Fig. 2) and behavioural adaptation (Stork 1987).

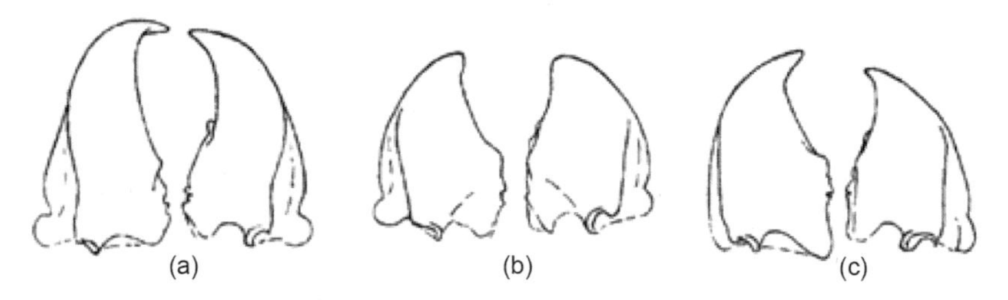

Fig. 2. Mandibles of (a) *Pterostichus adstrictus* Esch, (b) *Amara*, subg. *Curtonotus* and (c) *Harpalus aeneus* (After Carl H. Lindroth, *RES* 1974).

1.2 The External Anatomy of Ground Beetle

Body size of carabid beetle varies from 1.5 to 22 mm. Adults are voracious feeders, consuming close to their own body mass of food daily. Food is used to build fat reserves, especially before reproduction and hibernation (Thiele 1977). It is the largest Adephagan family and one of the most specious of beetle families. The suborder Adephaga is a relatively large group of specialized beetles that is morphologically defined by the presence of six abdominal ventrites, pygidial defence glands in the adult and liquid-feeding mouthparts in the larvae (Lawrence & Britton 1991). They are well proportioned cursorial beetles with prominent mandibles and palps, long slender legs, striate elytra, and sets of punctures with tactile setae. Most of the beetles have an antenna-cleaning organ and largely pubescent antennae. Adult ground beetles range in sizes from 2 mm to over 35 mm (about 1/8 inch to 1 ¼ inch). Many of the nocturnal species are dark black or brown; these are the ones that scurry away for cover when anyone turn over a dirt clod, rock, or log. Ground beetles can be distinguished from darkling beetles, which are also dark coloured and reside on the soil surface and by how fast they move. Diurnal (day-active) species tend to be iridescent and brightly coloured or patterned. Overall, the nocturnal species are larger than diurnal one. In hot countries, naturalism become more common; conversely, species that are nocturnal in central Europe became diurnal in the arctic. The adults typically have long legs, which allow them to move rapidly to capture prey and avoid other predators (Fig. 3).

They live in nearly every available habitat, although some species are associated with particular ecosystems, like meadows, woodlands, or crop fields. Due to the habitat specificity of some species, these beetles can be used as biological indicators to assess land use changes among different ecosystems. The beetles employ a wide variety of ecological strategies, however some generalizations can be made to represent the majority of species.

1.3 The Metamorphosis of Ground Beetle

Ground Beetles in the course of their life span undergo various changes in form of which the egg is superficially the most vulnerable of the life stages,

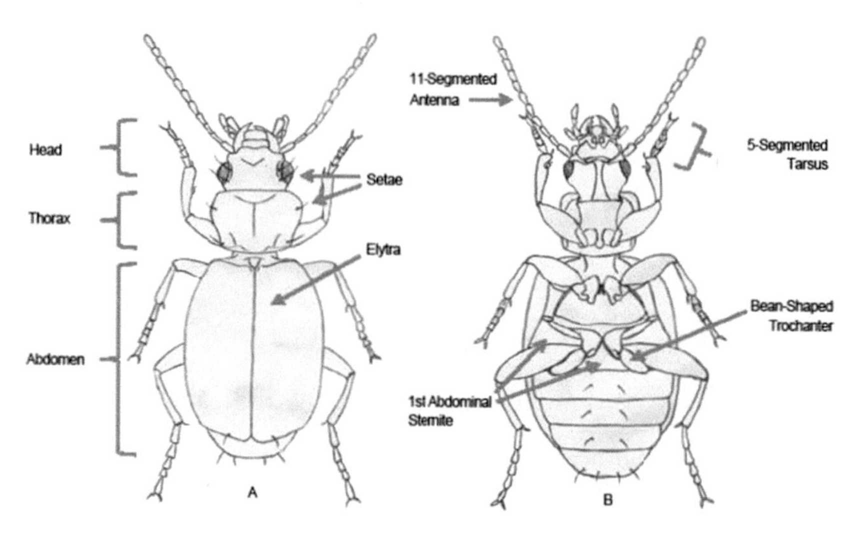

Fig. 3. (A) Dorsal and (B) ventral part of the ground beetle.
(*Source:* From https://extension.umaine.edu)

but ovipositing females can deliver eggs into microhabitats where their survival can be maximized. In carabids, female often have more prey type than males (Pollet & Desender 1987). Carabids are holometabolous insects that usually lay their eggs singly. Some species lay eggs in small or larger batches in crevices or in the soil after varying degrees of preparatory works by the female (Luff 1987, Thiele 1977). The female carefully choose the ovipositing site, sometimes excavating a chamber for the eggs. Some Pterostichini prepare a cocoon for a batch of eggs (Brandmayr & Zetto-Brandmayr 1979). Parental care, at its most developed, consist of no more than egg guarding or catching seeds in the egg chamber for the emerging larvae (Brandmayr & Zetto-Brandmayr 1979, Horne 1990). The typical carabid larva is free moving and campodiform (Crowson 1981) and usually undergo three stages before pupating in a specially constructed pupal chamber in the soil. Some species (for example *Harpalus* and *Amara* spp) have only two larval stages. Some tribes have specialized larvae with more larval stages that, in at least the later stage, exhibit reduced mobility. These species, which are ant or termite symbionts or specialized ectoparasites or predators (Erwin 1979b), total about 24% of all carabid tribes (Erwin 1979a). However, as not all members of these tribes exhibit these traits, these specialized larval bionomics characterize only a small minority of all species.

A greater diversity of food types in females are linked to greater egg size and egg number (Lovei and Sunderland 1996). Moreover, it was found that egg number and size are influenced by food consumption (Wallin et al. 1992). A study on egg production and body-mass changes indicate that field prey consumption by females allow them to realize 59% of their possible maximum egg production in May and 45% in June in Japan (Sota 1985).

Moreover, the egg stage was usually short, and the egg sacs contained the resources necessary for the completion of this life stage. The pupal stage is similarly sensitive. It lacks mobility and often lasts for longer periods, but is often better defended than the egg or larva. The larvae have limited mobility, weak chitinization, and therefore feeble tolerance to extremes, and it must also find sufficient food to develop (Figs 4 and 5). Larval feeding conditions often determins adult fertility as well (Nelemans et al. 1989). The larvae are campodeiform, have well-developed legs, antennae, and mandibles, and bear fixed urogomphi (Crowson 1981). Persistence in a habitat depend mostly on the life stage that is mostly vulnerable, as determined by the longest duration, narrowest tolerance limits, and most limited escape repertoire. All these factors point to the larval stage as the key to understanding occupation of habitat by a carabid species.

However, because larvae usually cannot migrate long distances, they have to survive in the environment where the egg-laying female abondoned them. Feeding conditions during larval development determined adult size, which is a major determinant of potential fecundity (Nelemans 1987a). The studies on the choice of habitat showed that the directed random walk, followed by a frequently turning walk in the presence of favourable conditions, will eventually lead carabids to their preferred habitats, but several different mechanisms helped beetles found or remain in suitable habitats. These mechanisms included internal clocks, sun-compass orientation (Colombini et al. 1994), and orientation either toward or away from silhouettes (Colombini et al. 1994, Rijnsdorp 1980). Some riparian ground beetles found their habitat by sensing volatile chemicals emited by blue algae present in the same habitat (Evans 1988). *Agonum quadripunctatum* (Buck.), a forest species in Europe and North America associated with burnt areas, is a good flyer and is probably attracted to the smell of smoke (Buakowski 1986). Some species (*Carabus nemarolis* Mueller) walked around in different habitat before settling in semi-natural habitats in preference to set-aside to arable areas (Kennedy 1994). The worldwide distribution of *Parena, Metallica, Pachycallida* and *Euproctinus* are shown in Fig. 6.

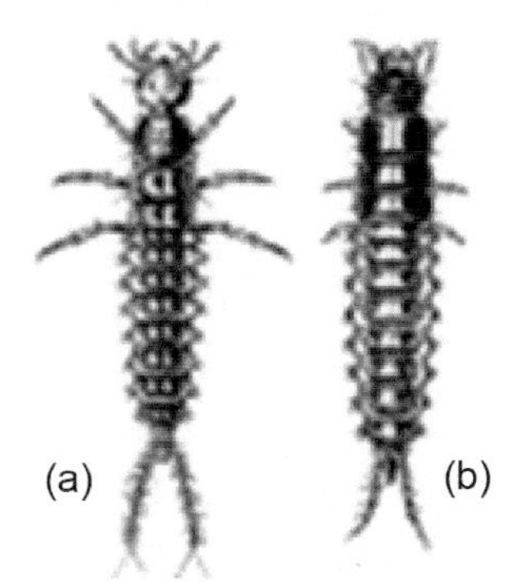

Fig. 4. Larvae. (a) *Nebria*; (b) *Agonum* (After Schiedte, redrawn.)

Fig. 5. Larva of beetle. (From https// en.m.wikipedia.org/wiki/File)

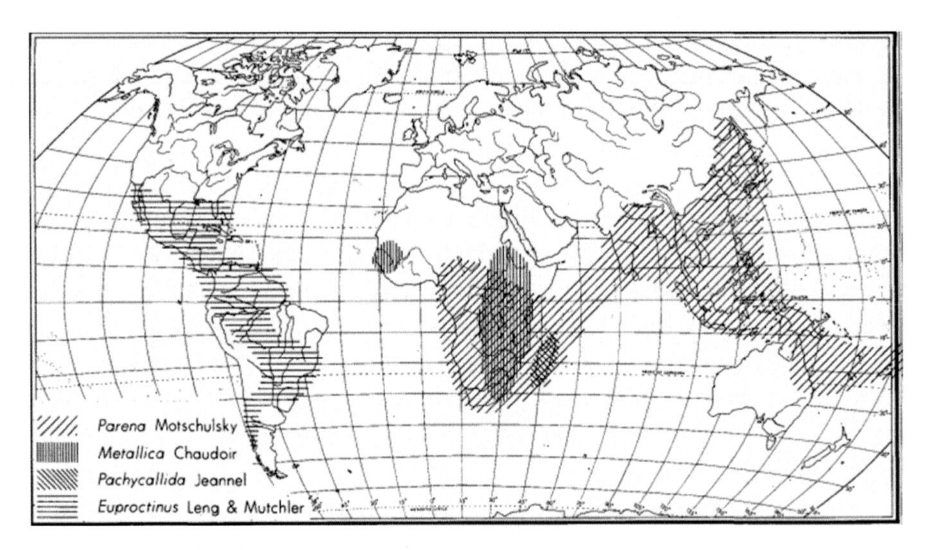

Fig. 6. Worldwide distribution of some important ground beetles. (http//creativecommons.org/liceses/by-nc-sa/3.0/us/)

1.4 Origin of Ground Beetles in Agricultural Ecosystems of Europe

Tischler (1958) reported that all or almost all of the ground beetles found in agricultural sites in western Europe originated from littoral habitats. The middle and upper levels of littoral habitats near the ocean and large rivers were rather unstable; wave or current actions constantly rework the shore producing small landslides. In this sense, the littoral habitat was similar to cultivated agricultural ones where the soil was physically reworked by the plough. Thus, it was no surprise that European species of ground beetles of agricultural habitats were not significantly affected by tillage. Carcamo et al. (1995) found no differences in the abundance of ground beetles in sites with tillage and without tillage, but found differences for some species. These beetles were pioneers in naturally disturbed and eroding soils near the Atlantic coast (Tischler 1958). Many of the ground beetles recorded in western Europe also occurred in eastern Europe where several steppe species were also found in agricultural sites.

1.5 Sources of European Species Introduced into North America

Lindroth (1957) discussed the most likely scenarios for the introduction of European species into colonial North America. The author also postulated that most species were introduced into Canada in boat ballasts where old shipledgers disclosing ships weighted down with barrels filled with soil obtained from the vicinity of the European ports of departure. Upon arrival in

North America, the ballasts were most likely emptied onto open habitats such as meadows or fields deforested for lumber production. Meadows, pastures, and lawns were excellent habitats for the non-native ground beetles, which were probably introduced independently on both the east and west coasts of North America. Once established, European species, most of which were capable of flight, could have spread further by invading weedy road edges and other open man-made habitats. Through train or truck commerce, European species could have moved great distances across Canada with agricultural products such as hay bales. In the prairie region, European species had generally been restricted to major cities such as Calgary, Edmonton, Regina, Saskatoon and Winnipeg. In these centres the survival of European species was favoured by such human interventions as watering lawns in summer (Spence and Spence 1988) or soils warmed by heated basements in winter.

1.6 Factors Affecting Distribution of Ground Beetle

Habitat and microhabitat distribution can be influenced by several factors as follows:

1. Presence and distribution of competitors. For example, forest carabids in Finland are influenced by the distribution of Formica ant species (Niemela 1990).
2. Life history and season. *Amara plebeja*, for instance, has different hibernation (Woodland) and reproduction (grassland) habitats. The beetles flow between habitats in spring and autumn. Flight muscles are temporarily autolysed between flights, then completely reconstructed for the return flight. In the autumn, they flew toward woodland silhouette shapes (van Huizen 1979).

As a group, carabids originally use fully functional wings as the primary dispersal mode. However, flight is very costly and is subjected to intense selection (Roff 1994). Once the benefits of flight do not match its cost, as on, for example, islands and mountain tops, it is quickly lost (Darlington 1943). Flightlessness and flight dimorphism have repeatedly evolved in carabids. For example, of the carabid fauna of Newfoundland (157 species), 12.7% are dimorphic and 21.0% flight-less, a condition reached through nine or more independent evolutionary transitions (Roff 1994). Environmental conditions may influence expression (Aukema 1991). Flight ability varies little between the sexes (Roff 1994). The proportion of flightless individuals in dimorphic species increase with increasing habitat persistency and time since colonization (den Boer et al. 1980). The proportion of macropterous *P. melanarius* can be as low as 2% in stable habitats (e.g. old forest patches) (den Boer 1970) or as high as 24-45% in less stable ones (e.g. newly reclaimed polder) (Haeck 1971).

Flight is greatly influenced by temperature, rain and wind (van Huizen 1979). In some species (such as the Palaearctic *Amara plebeja*), the flight muscles are broken down during egg production and then resynthesized; in others, flight capability during reproduction is not impaired, and up to 80% of dispersing females carry fertilized eggs. Females of more species from ephemeral habitats than from persistent ones carry ripe eggs (van Huizen 1990), which increases the probability of recolonizing empty habitat patches.

1.7 Assemblages of Ground Beetle

1.7.1 Ground Beetle in Semi-natural Habitats

Two wheat fields were chosen: one (FS) with an adjoining sown wildflower area (SWA) and other (FG) with an adjoining grassy margin. Altogether, 7070 carabid individuals belonging to 48 species were observed at Austria (see Appendix, Table 1). Nine hundred individuals of 34 species were found in SWA, 2319 individuals of 24 species were found in the fields adjoining SWA, 1454 individuals of 31 species were found in grassy margins, and 2397 individuals of 27 species were found in the fields adjoining grassy margins. SWA and the adjoining fields had 18 species in common and 22 species were shared between grassy margins and the adjoining fields. There were 21 species present in both types of semi-natural habitat (i.e. SWA and grassy margins) and 20 species were shared between wheat fields adjoining SWA and wheat fields adjoining grassy margins. By contrast to total species richness, there were significant differences in total activity-density among habitat types (Table 1). Activity-density was clearly higher in both FG and FS than in grassy margins and SWA, as indicated by non-overlapping notches (Fig. 7a). Carabid species were characterized by different ecological traits through their body size, wing size, nutrition and ecological preference (see Appendix, Table 1). Activity-density differed significantly among habitat types in all three body size classes (Table 1). There was a trend towards higher values in wheat fields compared with semi-natural habitats, which was particularly pronounced in medium-sized carabids (Fig. 7b).

In contrast to total species richness (Fig. 7a), richness of small carabids differed significantly among habitat types and with lowest in SWA (Fig. 7b). Activity-density of carabids with intermediate wing size and macropterous ones differed significantly among habitat types (Table 1), with a lower density of macropterous beetles in SWA and grassy margins compared with FS and FG (Fig. 7c). Species richness of beetles with intermediate wing size differed significantly among habitat types and is highest in grassy margins (Fig. 8c). Both activity-density and richness differed significantly among habitat types in polyphagous-carnivorous carabid beetles (Table 1). Densities of polyphagous-carnivorous beetles were considerably higher in fields compared with adjoining semi-natural habitats (Fig. 7d) and richness was higher in

Table 1. Analysis of variance showing effect of habitat type on total number of individuals (measured as activity-density) and species, and on individual and species numbers relative to the ecological traits body size, nutrition and ecological preference

Ecological traits	*Number of individuals*				*Number of species*			
	d.f	*ss*	*F*	*p (or p adjusted)*	*d.f*	*ss*	*F*	*p (or p adjusted)*
Total	20	20.74	9.14	1.9×10^{-13}***	20	112.17	1.68	0.055
Body size								
1-7 mm	20	5.64	0.28	0.0472*	20	20.19	1.01	0.0098054**
7.1-14 mm	20	12.52	0.63	$3,68 \times 10^{-14}$***	20	11.93	0.60	0.123475
>14 mm	10	4.20	0.42	0.0472*	10	1.53	1.75	0.2378383
Wing size								
Brachypterous	NA	NA	NA	NA	NA	NA	NA	NA
Intermediate	19	10.43	2.22	0.0131*	19	4.98	2.09	0.0438*
Macropterous	20	17.97	0.90	7.91×10^{-11}***	20	4.49	1.74	0.0663653
Nutrition								
Polyphagous herbivores	14	3.16	1.18	0.0620	14	1.55	1.90	0.1936825
Polyphagous carnivores	20	14.73	8.31	1.07×10^{-11}***	20	1.73	2.18	0.0402661*
Herbivores	NA	NA	NA	NA	NA	NA	NA	NA
Carnivores	15	3.48	0.23	0.158	15	1.31	1.61	0.198
Ecological preference								
Euryoecious	20	133.82	6.69	3.02×10^{10}***	20	12.86	1.34	0.545
Stenoecious	15	18.15	1.82	0.0927	15	4.36	1.01	0.692

p-value correction for multiple testing for each ecological trait was performed by the Benjamini-Yekutieli adjustment (i.e values shown in italic)
NA analysis of variance not available as a result of lack of data because of the small sample size. $P < 0.05$; $P < 0.01$; ***$P < 0.001$
(*Source:* Anjum-Zubir et al. 2015)

grassy margins than in SWA (Fig. 8d). Activity-density of euryoecious carabids differ significantly among habitat types and was considerably higher in FG and FS compared with SWA (Fig. 7e).

There were no significant differences in species richness among habitat types for ecological preference (Fig. 8e). When the number of recorded species was standardized to the number of individuals, species richness is by far highest in SWA. The number of species in SWA differed clearly from FS and FG for all sample sizes (Fig. 9). Because the 95% confidence intervals of the rarefaction curves of SWA and of FS and FG did not overlap, this indicated a significant difference between SWA and both FS and FG. Within the rarefaction curves, the steepest slope was observed for SWA which indicated that there were many species in SWA that were not recorded in the study. On the other hand, only few further species could be expected to occur in FG and grassy margins, and no further species were expected in FS, the habitat with the lowest species number.

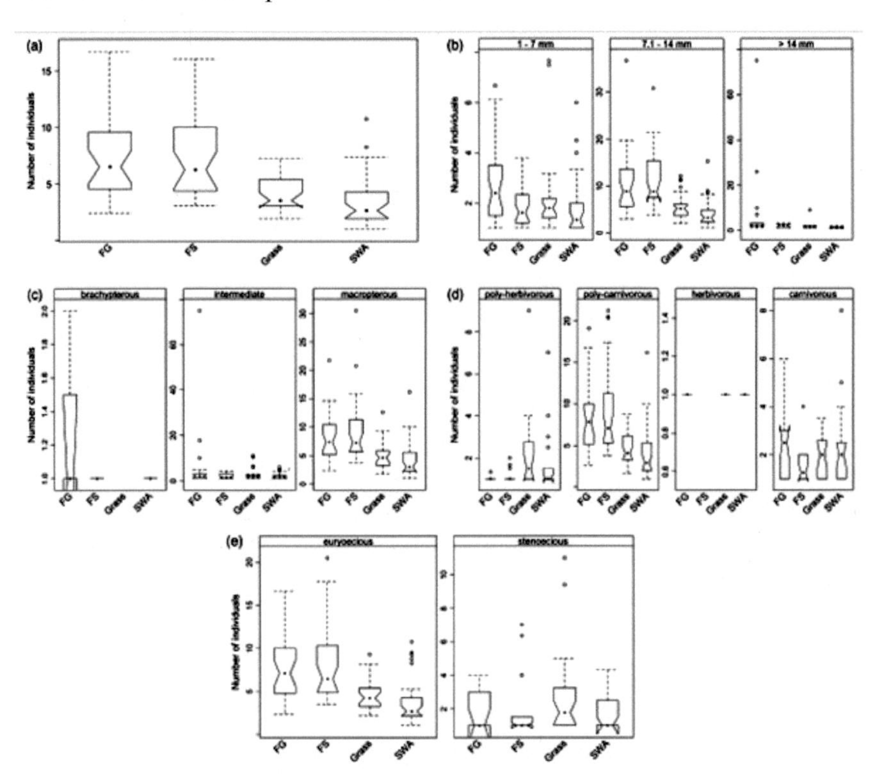

Fig. 7. Effect of habitat type on number of individuals measured as activity-density of ground beetles in wheat fields adjoining grassy margins(FG), wheat field adjoining SWA (FS), grassy margins (grass) and sown wildflower (SWA). Box-whisker plots show the medians (*), notches 25% and 75% percentiles, 10% and 90% percentiles (dashed line) and outlying values outside the percentiles (O). Non-overlapping notches indicated significant difference among habitat types: (a) total individuals, (b) body size, (c) wing size, (d) nutrition and (e) ecological preference. (*Courtesy:* Anjum-Zubair et al. 2015)

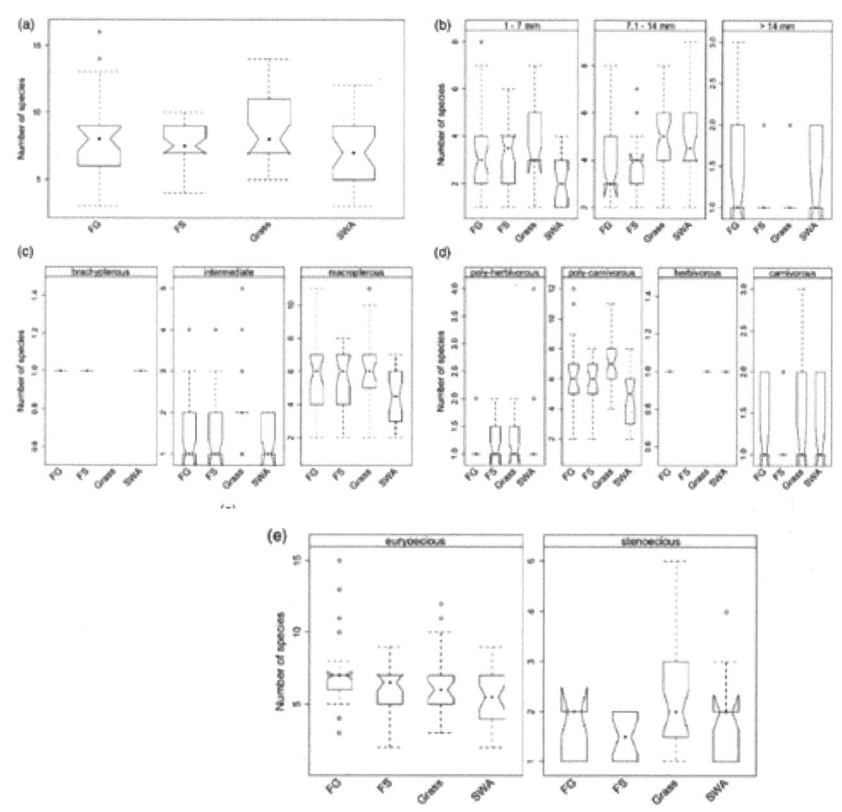

Fig. 8. Effect of habitat type on number of carabid species in wheat fields adjoining grassy margins (FG), wheat fields adjoining SWA (FS), grassy margins (grassy) and sown wildflower areas (SWA): (a) total species, (b) body size, (c) wing size, (d) nutrition and (e) ecological preference. (*Courtesy:* Anjum-Zubair et al. 2015)

To assess the effect of habitat type on carabid assemblage composition the trap count data are transformed by log ($x_{i+1)}$ and ordinate them with Non-metric Multidimensional Scaling (NMDS) with Bray-Curtis similarity measure. NMDS revealed a gradual change in carabid assemblage composition from wheat fields to sown wildflower areas (Fig. 10a). Semi-natural habitats on the one hand and wheat fields on the other hand were more similar to each other than to the directly adjoining habitat. Assemblages from FG and FS were indistinguishable. The grassy margins are a little more similar to the wheat fields than SWA. The right upper part of Fig. 10(b) revealed species typical for semi-natural habitats and the left lower part comprised of species preferably found in wheat fields.

The point-biserial correlation coefficient rpb indicates significantly positive associations for *Ophonus puncticeps* Stephens with SWA, *Bembidion properans* (Stephens), *Bembidion quadrimaculatum* Dejean, *Harpalus affinis* Schrank with grassy margins and *Anchomenus dorsalis* (Pont.) with FG (Table 2).

There were several species significantly and negatively associated with SWA (*Agonum muelleri* (Herbst.), *Loricera pilicornis* Fab, *Nebria brevicollis*

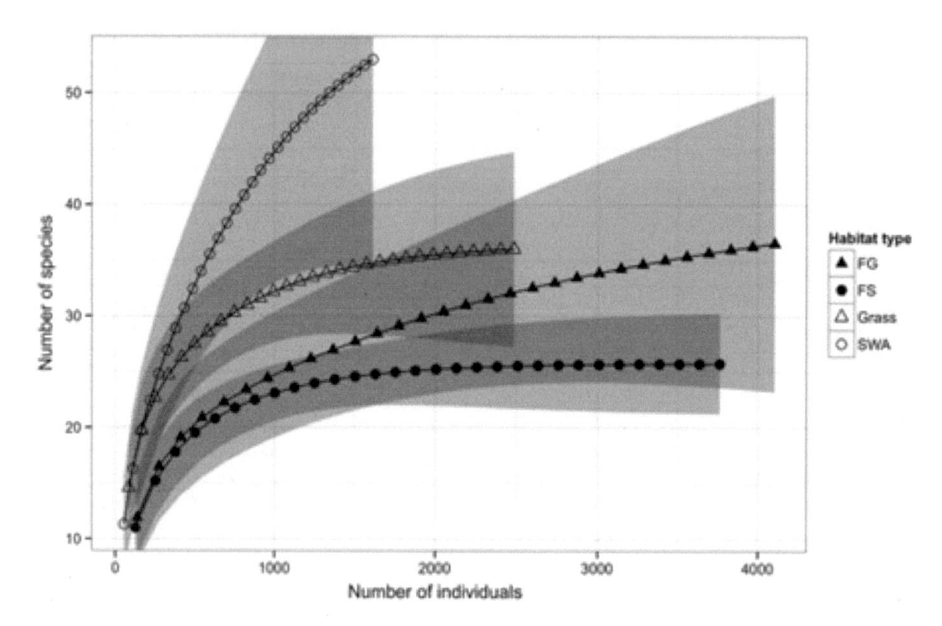

Fig. 9. Rarefaction curves showing the effect of habitat type on the number of species standardized for number of individuals sampled in each study site. The first 10 samples per curve were the recorded values, the additional 20 samples were the extrapolated values. Grey areas revealed 95% confidence intervals. FG, wheat fields adjoining grassy margins; FS, wheat fields adjoining SWA; Grass, grassy margins; SWA, sown wild flower areas. (*Courtesy:* Anjum-Zubair et al 2015)

(Fab.), *Poecilus cupreus* (Linn) and *Pterostichus melanarius* (Illiger)) and FS (*Anisodactylus binotatus* Fab). In the permanova model, which explained 47.8% of the variation in carabid assemblages, there was a highly significant effect of habitat type and a weaker significant effect of landscape, although there was no significant effect of field on the similarity of carabid assemblages (Table 3). Habitat type explained 25.6%, landscape explained 12.3%, and field explained 9.9% of total variation.

Ground beetles are considered as a beneficial component of agricultural systems: many species consumed insect pests and weed seeds. The cumulative effect of ground beetles on weeds and insect pests can be significant over time, given they can consume their own body weight in food daily. Ground beetles do not generally feed on plant matter other than seeds unless their diet can be supplemented due to scarcity of other food sources. They are rarely considered direct crop pests (https:// extension. umaine.edu.)

Carabid Numbers and Ecological Traits

In accordance with hypothesis 1, numbers of total carabid individuals were higher in the wheat fields adjoining grassy margins and SWA compared with both types of semi-natural habitats. Consistently, carabids were more abundant in wheat fields than in both adjoining grassy margins (Birkhofer et al. 2013) and meadows (Batary et al. 2012) and carabid activity was observed

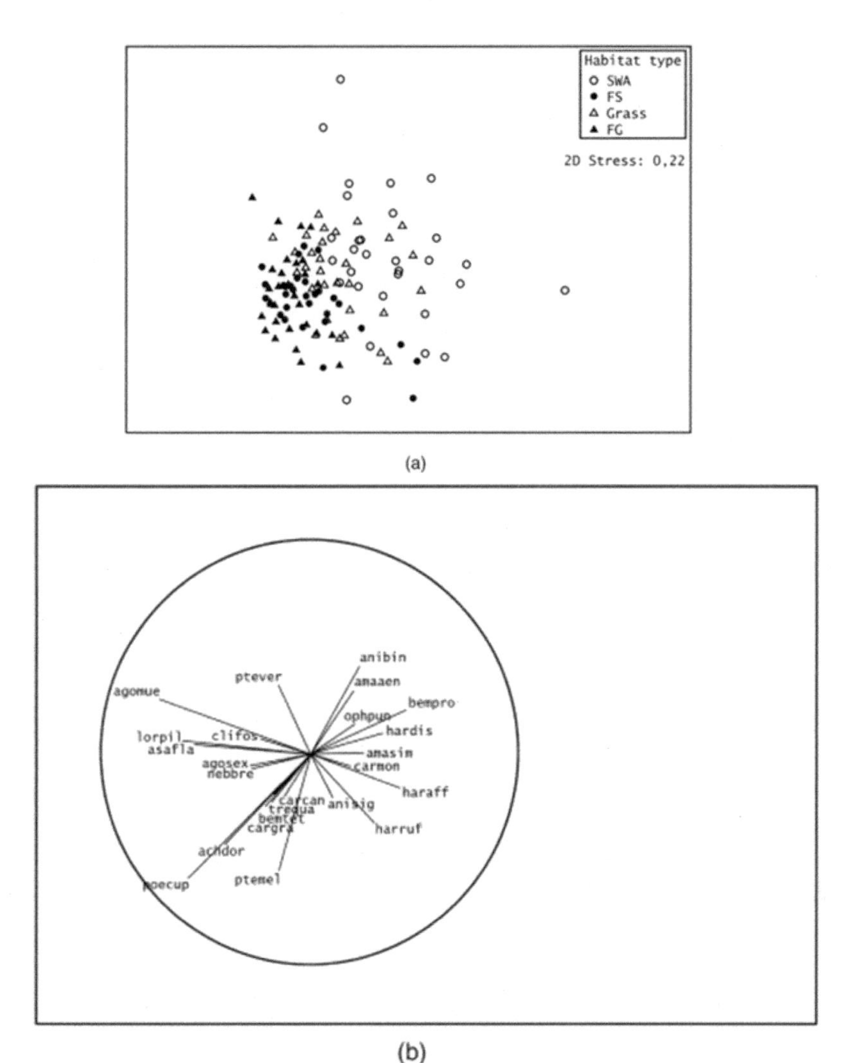

Fig. 10. Similarity of carabid assemblages in different habitat types: (a) Non metric multidimensional scaling ordination of carabid trapping positions sown in wildflower areas (SWA), fields adjoining SWA (FS), grassy margins (FG), stress is a goodness-of-fit measure for the relationship between the original resemblance matrix of the traps and its representation by the ordination. (b) The same ordination, with vector overlays for dominant species (Spearman correlation >0.2 between the species vectors and the ordination) 'agomul', *Agonum muelleri*; 'agosex', *Agonum sexpunctatum*; 'amaaen' *Amara aenea*; 'amasim' Amara similata; 'achdor', *Anchomenus dorsalis*; 'anbin', *Anisodactylus binotatus*; 'anisig', *Anisodactylus signatus*; 'asafla', *Asaphidion flavipes*; 'bempro', *Bembidion properans*; 'bemtet' *Bembidion tetracolum*; 'carcan', *Carabus cancellatum*; 'cargra', *Carabus granulates*; 'carmon', *Carabus monilis*; 'clifos', *Clivina fossor*; 'haraff', *Harpalus affinis*; 'hardis', *Harpalus distinguendus*; 'harruf', *Harpalus rufipes*; 'lorpal', *Loricera pilicornis*; 'nebbre', *Nebria brevicollis*; 'ophpun' *Ophonus punticeps*; 'poecup' *Poecilus cupreus*; 'ptemel' *Pterostichus melanarius*; 'ptever' *Pterostichus vernalis*; 'trequa' *Trechus quadristriatus*. (*Courtesy:* Anjum-Zubair et al. 2015)

Table 2. Association of carabid beetle species and four habitat types (r_{pb}) association index, lower and upper confidence intervals in bracket

Species	*FG*	*FS*	*Grass*	*SWA*
Agonum mueleri (Herbst.)				-0.46(-0.62,-0.30)
Anchomenus dorsalis (Pont.)	0.51(0.24,0.74)			
Anisodactylus binotatus Fab.		-0.38(-0.53,-0.21)		
Bembidion properans (Stephens)			0.58 (0.29, 0.76)	
Bembidion quadrimaculatum Dejean			0.58 (0.22,0.83)	
Harpalus affinis Schrank			0.42(0.04,0.70)	
Loricera policornis Fab.				-0.32(-0.47;-0.20)
Nebria brevicollis (Fab.)				-0.35 (-0.49,-0.23)
Ophonus puncticeps Stephens				0.48 (0.00,0.71)
Poecilus cupreus (Linn.)				-0,42(-0.57,-0.27)
Pterostichus melanarius (Illiger)				-0.14(-0.48,-0.10)

Negative values indicate negative associations and vice versa. For clarity, only significant associations are shown
FG, wheatfields adjoining grassy margins, FS, wheatfield adjoining SWA, Grass, grassy margins; SWA, sown wildflower association.
(*Courtesy:* Anjum-Zubair Mohammed 2015)

Table 3. Permutational multivariate analysis of variance showing the effect of landscape, field and habitat type of similarity of carabid assemblage

Source	*df*	*MS*	*P*	*Perms*
Landscape	9	5024.3	0.0184	9871
Field (landscape)	10	3221.7	0.1763	9870
Habitat type (field landscape)	20	2631.8	0.0001	9803

Perms: number of permutations

to be higher in intensively managed meadows compared with adjoining extensive ecological compensation meadows (Albrecht et al. 2010). These patterns might be a result of increased plant productivity at higher fertilization levels which potentially provides more prey for predatory insects (Siemann 1998; Haddad et al. 2000) or, may be more likely, a result of spillover. When *P. cupreus* was released at a junction of wheat fields and grassy margin strips, beetles move directly to wheat centres (Ranjha & Irmler 2014). Because, in this study, *P. cupreus* was by far the most abundant carabid, its aforementioned movement pattern can help to explain why higher carabid activity-density are found in intensively managed wheat compared with less disturbed semi-natural habitats. Also, the finding is most likely the result of a more densely covered soil surface impeding movement of carabids in semi-natural habitats (Honek and Jarosik 2000).

Total carabid species richness differed between habitat types only with marginal significance ($P = 0.055$), thus providing only limited support for the expected positive effects of semi-natural habitats on carabid richness. These richness patterns might be a result of the limited sampling period. Similarly, species richness of carabid beetles did not significantly differ among grassy margins and adjoining wheat interiors (Birkhofer et al. 2013) and grassy margins were observed to not necessarily present areas of particularly high animal biodiversity compared with arable fields (Ernoult et al. 2013). Besides several studies revealing higher carabid species numbers in semi-natural habitats compared with adjoining arable fields (Luka et al. 2001, Yu et al. 2006, Werling and Gratton 2008), there are studies observing similar species numbers in semi-natural habitats and arable fields (Saska et al. 2007, Albrecht et al. 2010) showing that wheat fields are appropriate habitats for many carabid species. Thus, there are contradictory observations on carabid.

Indicators of diversity (population density, relative number of species and evenness) in ground beetles (Carabidae) and rove beetles (Staphylinidae) in upland rice fields were assessed between 1995 and 1999 at Garoua in the Benue valley in North Cameroon. A total of 4369 beetles belonging to 45 carabid species and 2109 beetles belonging to 31 staphylinid species were caught in pitfall traps. Among the carabid beetles, five species *Scarites (Orientolobus) lucidus strigiceps* Quedenfeldt, *Chlaeniostenus denticulatus elatus* (Erichson), *Lissauchenius venator* (LaFerté), *Pheropsophus marginatus*

(Dejean) and *Abacetus crenulatus* Dejean, in decreasing order, were dominant. The Shannon-Weiner and Evenness indices varied slightly from year to year. Diversity values remain relatively low among the staphylinid beetles, revealing that rice fields were under-populated by this group of polyphagous predators. The role of predatory soil-surface beetles are similar to that of an IPM component in West African rice ecosystems (Woin et al. 2005). Some species recorded as predator of crop pests in India (Fletcher 1919, Puttarudriah and Raju 1952, Samaland Misra 1978, Prasadkumar and Rajagopal 1990). They are distributed throughout the world and in all the three Zoogeographical region of India viz Himalayan tracts, Plains and Decans (Andrewes 1929).

1.7.2 Ground Beetle (Coleoptera: Carabidae) Assemblages in a Transgenic Corn–Soybean Cropping System

As generalist predators, ground beetles can play an important role in keeping primary and secondary pest populations below economic thresholds (Potts and Vickerman 1974, Landis et al. 2000). The objectives of this study are (1) to ascertain the general species composition of ground beetle assemblages in southeastern South Dakota, (2) determine differences in species composition of ground beetles between crops in a Bt corn-soybean (*Glycine max* Merrill) cropping system, (3) describe the temporal structure of ground beetle assemblages in a Bt corn-soybean cropping system, and (4) describe the spatial structure of ground beetle assemblages in a Bt corn-soybean cropping system. This study was conducted in 2000, 2001, and 2002 in Brookings County, SD, on four fields of rotated corn and soybean in Aurora Township of USA. Two of the fields were started in corn and the remaining two in soybean. The fields ranged in size from 16.1 to 16.5 ha (~200 × 800 m) and had been in a corn-soybean rotation for several years before the study. On the north and south side of fields 1, 2 and 3, and on the north side of field 4, a single row (~6.5 m wide) of Ponderosa pine trees, *Pinus ponderosa* Lawson, served as a windbreak. A 200 × 800-m field in corn-soybean rotation was on the north side of field 1, and a multiple row windbreak (~30 × 800 m) was on the south side of field 4. A multiple row windbreak (~200 × 230 m) near the house (Fig. 11) bordered a portion of the south side of field 1 and the north side of field 2. A grassy fence-row bordered the fields on the east, and a grassy roadside ditch bordered the fields on the west. Two east-to-west transects ~60-80 m from the south (A transects) and north (B transects) boundaries along each of the four fields. Fourteen pitfall traps were placed ~60 m apart in each transect to capture ground beetles (Fig. 11).Trap design followed that of Morrill (1975) and consisted of a 455-ml Solo cup (Concept Communications, Burr Ridge, IL) with a 145-mm i.d., a Solo Cozy Cup funnel, and an inner 148-ml Solo cup partially filled with propylene glycol as a preservative. Pitfall traps were set in July of each year. Traps are opened for 48 h each week through September (August in 2000). This sampling period covered the peak activity and most of the activity period for ground beetles in this area (Kirk 1971).

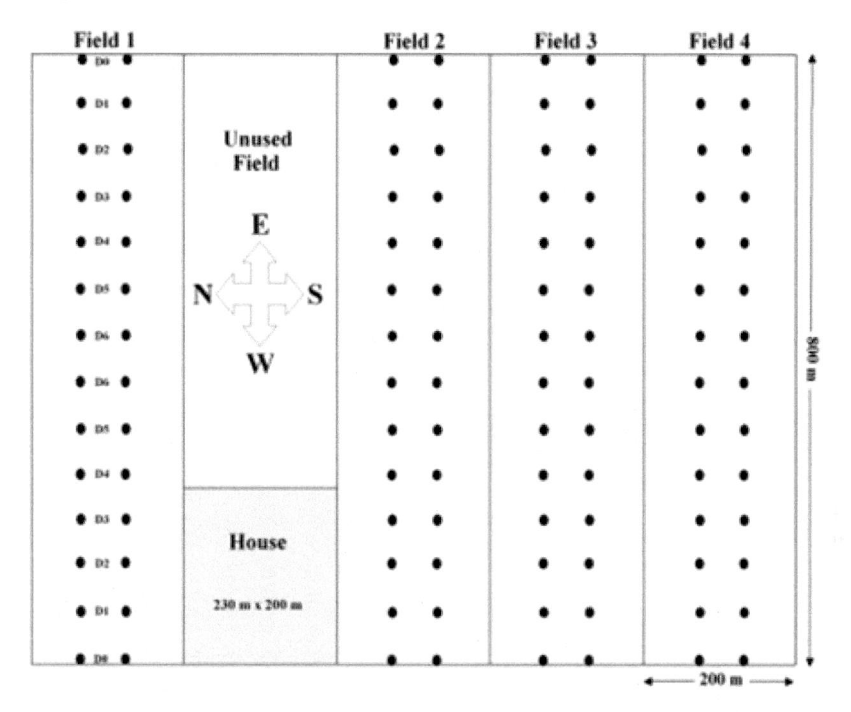

Fig. 11. Arrangement of fields and pitfall traps. Two east-to-west transects ~60-80 m from the south (A transects) and north (B transects) boundaries were established along each of the four fields. Traps were placed at ~60 m intervals from the borders. The borders represented an abrupt change in vegetation from crop to grass. Distance from border; D_0 = 0 m, D_1 = 60 m, D_2 = 120 m, D_3 = 180 m, D_4 = 240 m. D_5 = 300 m, and D = 360 m (*Source:* French et al. 2004)

Species Data

Overall in three years, 24,750 ground beetles were captured, representing 57 species (Table 4). The beetles captured ranged in size from 2 mm for *Elaphropus anceps* (LeConte) and *Dyschirius globulosus* (Say) to 25 mm for *Calosoma calidum* (F.), *Harpalus caliginosus* (F.), and *Scarites subterraneus* (F.). Of the 57 species collected, three [*Cyclotrachelus alternans* (Casey), *Harpalus pensylvanicus* (DeGeer) and *Pterostichus permundus* (Say)] accounted for 81% of all individuals captured dominant species. Six other species [*Bembidion quadrimaculatum* Say, *Brachinus ovipennis* LeConte, *Calosoma calidum* (F.), *Cicindela punctulata* Olivier, *Poecilus chalcites* (Say) and *P. lucublandus* (Say)], accounted for an additional 14% of all individuals captured (Table 4), which were considered to be abundant species.

If a total of 50-175 individuals of a species were collected, they were considered common. Rare species accounted for 38 of the 57 captured, and 15 species were captured only once (Table 4). There was no significant interactions between crop and distance for the total number of beetles captured, each of the dominant species, or the six abundant species captured. Across all species,

Table 4. Number and percentage of beetles captured for each species and abbreviations (Abbrev^{n}) of species depicted in biplots of canonical correspondence analysis

Abbrev^{n}	*Species*	*No*	*%*	*Abbrev^{n}*	*Species*	*No*	*%*
Acp	*Acupalpus pauperculus* Dejean	3	<0.01	Cip	*Cicindela punctulata* Oliver	942	3.5
Agc	*Agonum cupripenne* (Say)	12	<0.01	Cir	*Cicindela repunda* Dejean	1	<0.1
Agp	*Agonum placidum* (Say)	91	0.4	Clb	*Clivina bipunctulata* (F.)	30	0.1
Ama	*Amara angustata* (Say)	1	<0.1	Cli	*C. impressefrons* LeConte	5	<0.1
Amc	*A. carinata* (Le Conte)	91	0.4	Cya	*Cyclotrachelus alternas* (Casey)	8,949	36.2
Ame	*A. exarata* Dejean	4	<0.1	Cyp	*Cymindis pilosus* Say	2	<0.1
Ami	*A. impuncticollis* (Say)	1	<0.1	Dip	*Discoderus parallelus* (Hald.)	3	<0.1
Aml	*A. latior* (Kirby)	2	<0.1	Dyg	*Dyschirius globulosus* (Say)	2	<0.1
Ali	*A . littoralis* Mannerheim	1	<0.1	Ela	*Elaphropus anceps* (Le Conte)	96	0.4
Amo	*A. obesa* (Say)	33	0.1	Gaj	*Galerita janus* (F.)	2	<0.1
Anh	*Anisodactylus harrisii* Le Conte	1	<0.1	Hac	*Harpalus calignosus* Fab.	60	0.2
Anr	*A. rusticus* (Say)	36	0.1	Hae	*H. erraticus* Say	103	0.4
Ans	*A. sanctaccrucis* (F.)	5	<0.1	Haf	*H. faunus* Say	3	<0.1
Ban	*Badister notatus* Hald.	1	<0.1	Hah	*H. herbivagus* Say	55	0.2
Bem	*Bembidion mimus* Hayward	3	<0.1	Hao	*H. opacipennis* (Hald.)	1	< 0.1
Beq	*B. quadrimaculatum* Say	891	3.6	Hap	*H. pensylvanicus* (DeGeer)	6,583	26.6
Ber	*B. rapidum* (Le Conte)	22	<0.1	Hav	*H. ventralis* Le Conte	10	<0.1
Brj	*B. janthinipennis* (Dejean)	1	<0.1	Min	*Microlestes nigrinus* (Manner.)	159	0.6

(Contd.)

Bro	*B. ocipennis* Le Conte	244	1.0	Poc	*Poecilus chalcites* (Say)	425	1.7
Brq	*B. quadripennis* Dejean	2	<0.1	Pol	*P. lucublendus* (Say)	937	3.8
Cag	*Calathus gregarious* (Say)	99	0.4	Por	*Polyderis rufotestacea* (Hay)	2	<0.1
Cac	*Calosoma calidum* (F.)	1 56	0.8	Ptc	*Pterostichus coracinus* (New)	1	<0.1
Cao	*C. obsoletum* Say	1	<0.1	Ptf	*P. femoralis* (Kirby)	11	<0.1
Cas	*Carabus serratus* Say	1	<0.1	Ptm	*P. melanarius* (Illiger)	3	<0.1
Che	*Chalaenius emarginatus* Say	1	<0.1	Ptp	*P. permundus* (Say)	4,450	18.0
Chp	C. *platyderus* Chaudoir	53	0.2	Scs	*Scarites subterraneus* F.	101	0.4
Chs	*C. sericeus* (Forster)	9	<0.1	Stc	S. *comma* (Fab.)	1	<0.1
Cht	*C. tomentosus* (Say)	6	<0.1	Syi	*Synuchus impuctatus*(Say)	1	<0.1
Ctr	C. *tricolor*	Dejean	1	<0.1			

Source: French et al. 2004

although more beetles were captured in the corn fields than in the soybean fields, the difference was not significant (Table 5).

Table 5. Least square mean ± SE relative abundances per trap ($n = 168$ for each crop overall 3 yrs) for species of ground beetles captured during 2000-2002 in field of Bt corn and soybean

Species	*Crop*		*Anova*		
	Corn	*Soybean*	*F*	*df*	*p*
All species	78.36±37.05	68.90± 37.05	0.50	15.37	0.511
B. quadrimaculatum	2.72± 0.74	2.58± 0.74	0.04	15.22	0.854
B. ovipennis	0.58± 0.27	0.87 ± 0.27	1.10	1,8	0.325
C. calidum	0.13± 0.33	0.99 ± 0.33	16.88	15.61	<0.01
C. punctulata	2.80± 1.70	2.80 ± 1.70	0.21	16.46	0.664
C. alternans	30.47± 15.57	22.80 ± 15.57	2.62	15.1	0.165
H. pensylvanicus	20.40± 14.15	18.79± 14.15	2.66	15.31	0.160
P. chalcites	1.17± 0.66	1.39 ± 0.66	1.09	15.79	0.338
P. lucublandus	2.74± 0.88	2.84± 0.88	3.31	14.39	0.137
P. permudus	14.92± 6.58	11.57± 6.58	0.38	15.66	0.563

Source: French et al. 2004

The three dominant species, *C. alternans*, *H. pensylvanicus* and *P. permundus*, were also captured more often in the cornfields, but the differences were not significant (Table 5). There were no signicant differences in numbers captured in cornfelds and soybean fields for the other species (*B. quadrimaculatum*, *C. punctulata, P. chalcites* and *P. lucublandus*) except for *C. calidum*, which were captured predominantly in soybean fields. Overall species showed significant differences in the numbers of beetles captured with respect to distance into the fields of Bt corn and soybean from the field edges ($F = 14.46$, $df = 6{,}300$, $P > 0.0001$ (Table 6).

The greatest numbers of ground beetles were captured at the field borders. For each distance from 60 to 360 m into the fields, the numbers captured differed significantly from the numbers captured in the border. Similarly, there also were significant differences in numbers captured for each of the three dominant species, *C. alternans* ($F = 7.10$, $df = 6{,}300$, $P > 0.0001$), *H. pensylvanicus* ($F = 3.84$, $df = 6{,}300$, $P > 0.0011$), and *P. permundus* ($F = 6.63$, $df = 6{,}300$, $P > 0.0001$), because they each were captured most often near the field borders (Table 6). There were, however, substantial numbers of all three species captured at each distance into the fields. Of the six abundant species, only *C. punctulata* ($F = 5.06$, $df = 6{,}300$, $P > 0.0001$) and *P. chalcites* ($F = 3.24$, $df = 6{,}300$, $P > 0.0043$) showed significant differences in numbers captured with respect to distance from the borders. However, these species also were captured in substantial numbers at each distance into the fields. Indeed, a multiple comparison test for *P. chalcites* indicated no

Table 6. Least square mean ± SE relative abundances per distance from field bounders (n = 48 for each crop overall 3 yrs) for species of ground beetles captured during 2000-2002 in field of Bt corn and soybean

Species	*Approximate distance from field border (m)*						
	0	*60*	*120*	*180*	*240*	*300*	*360*
All species	$105.87\pm36.00_a$	$75.42\pm36_b$	$65.73\pm36.00_b$	$61.54\pm36_b$	$66.29\pm36._b$	$70.35\pm36_b$	$70.19\pm36_b$
B. quadrimaculatum	$3.06\pm0.76_a$	$2.33\pm0.76_b$	$2.40\pm0.76_a$	$2.54\pm0.76_a$	$2.54\pm0.76_a$	2.81 ± 0.76^a	$2.54\pm0.76_a$
B. ovipennis	$0.98\pm0.28_a$	$0.60\pm28_b$	$0.54\pm0.28_a$	$0.75\pm0.28_a$	$0.75\pm0.28_a$	$0.75\pm0.28_a$	$0.96\pm0.28_a$
C. calidum	$0.44\pm0.35_a$	$0.46\pm0.35_b$	$0.63\pm0.35_a$	$0.38\pm0.35_a$	$0.38\pm0.35_a$	$0.77\pm0.35_a$	$0.60\pm0.35_a$
C. punctulata	5.60±1.51a	$3.23\pm1.51_b$	$2.52\pm1.51_b$	$1.92\pm1.51_b$	$1.92\pm1.31_b$	$2.48\pm1.51_b$	$1.90\pm1.51_b$
C. alternans	$37.00\pm15.26_a$	$30.27\pm15.26_b$	$24.54\pm15.26_b$	$21.60\pm15.26_b$	$24.04\pm15.26_b$	$24.71\pm15.26_b$	$24.27\pm15.26_b$
H. pensylvanicus	$25.15\pm13.64_a$	$19.08\pm13.64_b$	$16.67\pm13.64_b$	$17.23\pm13.64_b$	$18.31\pm13.64_b$	$19.56\pm13.64_b$	$21.15\pm13.64_b$
P. chalcites	$1.73\pm0.70_a$	$2.00\pm0.70_a$	$1.77\pm0.70_a$	$1.10\pm0.70_a$	$0.69\pm0.70_b$	$0.69\pm0.70_b$	$0.88\pm0.70_a$
P. lucublandus	$3.85\pm0.95_a$	$2.58\pm0.95_a$	$2.54\pm0.95_a$	$2.06\pm0.95_a$	$2.75\pm0.95_a$	$2.60\pm0.95_a$	$3.13\pm0.95_a$
P. permundus	$20.02\pm6.3_a$	$11.92\pm6.36_b$	$12.00\pm6.36_b$	$11.05\pm6.36_b$	$11.58\pm6.36_b$	$13.94\pm6.36_b$	$12.17\pm6.36_b$

Mean comparisons were tested against DO using t-tests with Dunnet adjustments for type 1 errors, experiment-wise error = 0.05
(Source: French et al. 2004)

significant differences with respect to distance from the border (Table 6). There were no significant differences in numbers captured with respect to distance for *B. quadrimaculatum, B. ovipennis, C. calidum* and *P. lucublandus* (Multivariate Analysis). The Eigen values of the canonical correspondence analysis measure the proportion of total variation in ground beetle relative abundance explained by each respective axis (ter Braak 1986, 1995, ter Braak and Šmilauer 1998). The Eigen values, based on species relative abundances, for axis 1-4 were 0.101, 0.070, 0.048 and 0.024. Axis 1 accounted for 31.5% of the species-environment relationship, and together with axis 2, accounted for 53.3% of the species-environment relationship accounted for 76.0% of the total species-environment relationship. A biplot of the most important environmental variables and species scores illustrates that axis 1 represents an annual gradient (Fig. 12).

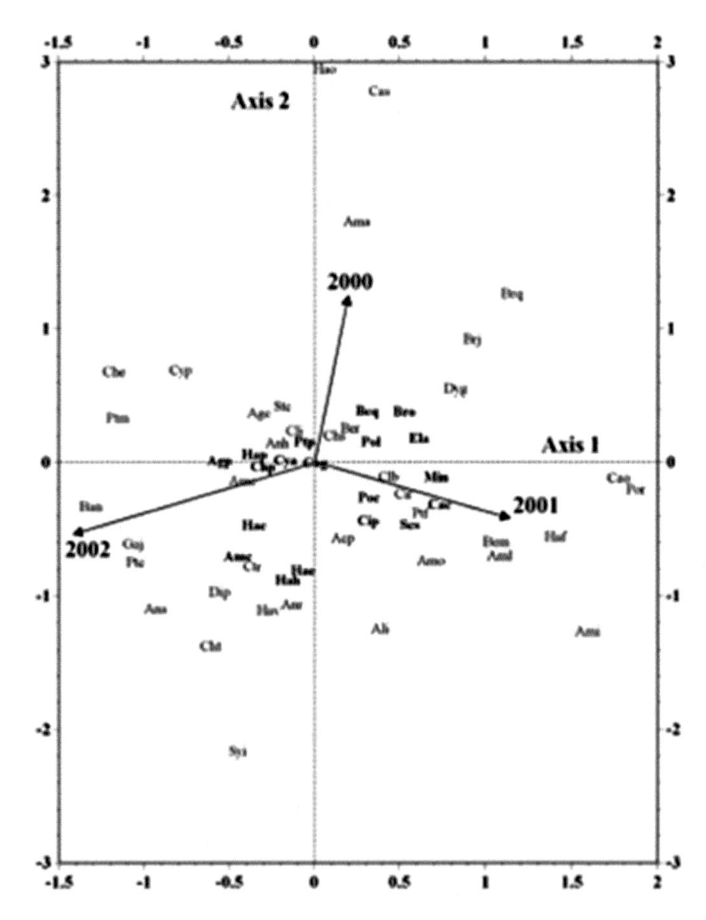

Fig. 12. Biplot of ground beetle relative abundances and most important environmental variables from the canonical correspondence analysis. The abbreviations of species names are plotted, and complete names are listed in Table 4. The 19 dominant, abundant, and common species are shown in bold. Arrows represent environmental variable. (*Source:* French et al. 2004)

The three dominant, six abundant, and 10 common species are depicted in bold. Species names and abbreviations are given in Table 4. Environmental variables are represented by arrows, and a relatively long arrow positioned close to an axis indicates a strong relationship with that axis (ter Braak 1986, Palmer 1993), such as years 2001 and 2002 with axis 1 (Fig. 12). Ground beetles positioned close to the arrows had a strong association with that variable, such as *C. calidum* (Cac) and year 2001, whereas ground beetles occurring near the origin of the axis represented perennial species that were captured in relatively similar numbers in all three years [e.g., *C. alternans* (Cya), *C. gregarius* (Cag), *C. platyderus* (Chp) and *P. permundus* (Ptp)]. Beetle assemblages that predominated in 2001 had positive values on axis 1 and ordinated to the right of axis 2. Beetle assemblages that predominated in 2002 had negative values on axis 1 and ordinated to the left of axis 2. The ground beetle assemblage associated with 2000 had positive values on axis 2 and ordinated above axis 1. The observed patterns for ground beetles with environmental variables were significantly different from random (Monte Carlo test statistic =3.25, P> 0.005) (ter Braak and Sˇmilauer 1998). Partial canonical correspondence analysis was used to depict the effects of crop types and distance from field borders on patterns of species relative abundance. In this partial analysis, the effects on species composition of years and fields were factored out as covariables. The Eigen values for axis 1-4 were 0.049, 0.039, 0.011 and 0.006, respectively. Again, these values measured the amount of variation in species scores explained by their respective axis, with axis 1 explaining more variation in species scores than axis 2 or 3. Of the variation in species composition remaining after factoring out the covariables, axis 1 accounted for 40.6% of the species-environment relationship, and together with axis 2, accounted for 73.5% of the species-environment relationship. A biplot of the environmental variables and species scores revealed that crop types were closely associated with axis 1 (Fig. 13).

The second axis separated ground beetle species which occupied field edges from those occupying field interiors. Ground beetles occurred near the origin of the axis might represent habitat generalists [e.g., *B. quadrimaculatum* (Beq), *B. ovipennis* (Bro), *C. punctulata* (Cip), *C. alternans* (Cya) and *E. anceps* (Ela)], whereas species occurred far from the origin represented crop or edge specialists [e.g., *M. nigrinus* (Min) and *H.erraticus* (Hae)]. Species associated with soybeans ordinated in the positive space of axis 1 and ordinated to the right of axis 2 [e.g., *C. calidum* (Cac)], whereas species associated with corn ordinated in the negative space of axis 1 and ordinated to the left of axis 2 [e.g., *H. pensylvanicus* (Hap)]. Species of ground beetles associated with field edges of both cornfields and soybean fields ordinated in the positive space of axis 2 [e.g., *C. gregarius* (Cag)]. Increasing distance from 60 to 360 m had little effect on ground beetle assemblages. The observed patterns for ground beetles with environmental variables were significantly different from random (Monte Carlo test statistic = 2.09, $P > 0.005$) (ter Braak and Šmilauer 1998).

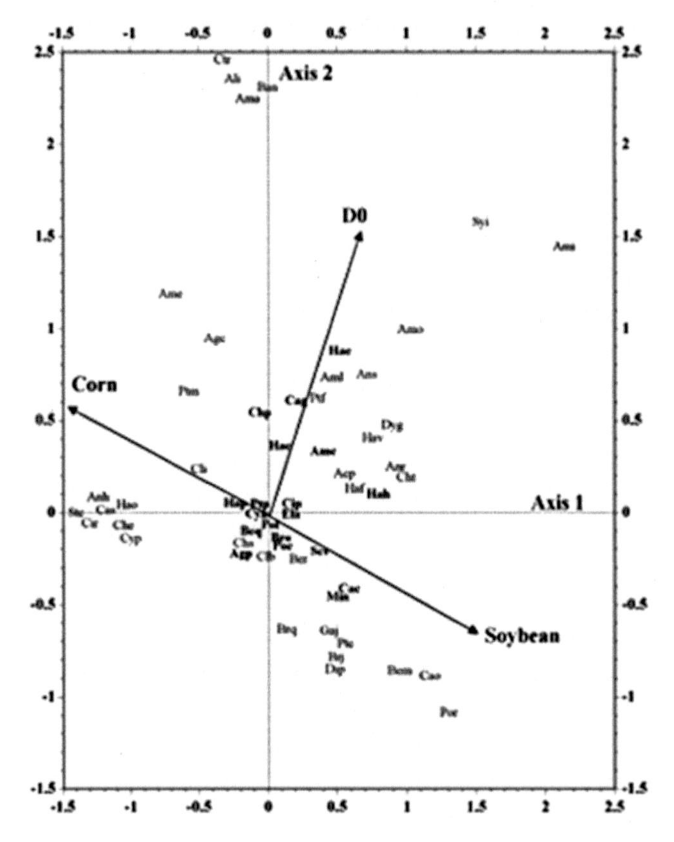

Fig. 13. Biplot of ground beetle relative abundances and environmental variables from a partial canonical correspondence analysis. The abbreviation of species names are plotted and complete names are listed in Table 4. The 19 dominant, abundant and common species are shown in bold. Arrows represent environmental variable. (*Source:* French et al. 2004)

It is typical for a few species to dominate ground beetle assemblages in terms of relative abundance and to vary in numbers over space and time (Thiele 1977, Luff 2002). In this study, three species, *C. alternans*, *H. pensylvanicus* and *P. permundus* (5.3%), of the 57 species collected accounted for 81% of all ground beetles captured. Six other species accounted for an additional 14%. Other studies also had found that a few species dominated the ground beetle fauna in agroecosystems (Kirk 1971, Barney and Pass 1986, Laub and Luna 1992, Tonhasca 1993, Ca′rcamo 1995, Ellsbury et al. 1998, French et al. 1998, French and Elliott 1999a, b). Although numbers captured differed from this study, the dominant and common species captured by Kirk (1971) and Ellsbury et al. (1998) in South Dakota were similar to those captured in the experiment. Both studies reported *C. alternans* as a dominant or most commonly found species. In contrast, Ellsbury et al. (1998) reported it to be a dominant species in soybean fields (House and All 1981, Wiedenmann et al. 1992) and cornfields (Esau 1975, Best et al. 1981). In contrast, among the abundant species captured, only *C. calidum* was captured in higher numbers in soybean fields than in cornfields. The factors determining habitat preference for *C. calidum* were unknown but might be due in part to differences in prey availability between cornfields and soybean fields. For example, *C. calidum* belonged to a genus commonly called "caterpillar hunters" because they tended to prey on lepidopteran larvae (Thiele 1977, Toft and Bilde 2002), and because the Cry3A1 toxin targets lepidopterans. Subsequent reduction in lepidopteran larvae might cause *Calo*soma species to seek prey elsewhere, such as soybean fields or other habitats.

Crop type had no impact on *B. quadrimaculatum, C. punctulata, P. chalcites* and *P. lucublandus* because they were captured in equal numbers in the cornfields and soybean fields. For ground beetles to be effective biological control agents against agricultural pests, they must be able to occupy the centre of the arable fields (Wissinger 1997, Landis et al. 2000). The highest captured number of beetles occurred at the field borders; however, corresponding numbers of beetles were captured from 60 to 360 m into the fields. The three dominant species, *C. alternans, H. pensylvanicus* and *P. permundus*, also were captured most frequently at the field borders; however, they too were captured in substantial numbers at field centres. In Iowa, Esau and Peters (1975) and Best et al. (1981) found *H. pensylvanicus* predominantly in field edges, but they also captured many within cornfields. French and Elliott (1999a, b) captured *H. pensylvanicus* most often in natural habitats and their borders rather than in wheat fields. *B. quadrimaculatum, B. ovipennis, P. chalcites* and *P. lucublandus* were captured equally throughout the fields, and because they were captured in equal numbers in the crop fields, probably represented habitat generalists. The distribution of *P. chalcites* varied only slightly from the borders into the field centres. In contrast to *H. pensylvanicus* (French and Elliott, 1999a, b) *P. chalcites* was captured most often in wheat field interiors rather than in natural habitats and field borders. French and Elliott also regarded *P. chalcites* as a synanthropic species, meaning it shared a close association with human activities and had probably benefited from agriculture in general (Spence and Spence 1988). *C. calidum* also were captured in equal numbers throughout the fields; however, they were more abundant in soybean fields. *C. punctulata* varied in numbers captured from the field borders to field centres. The distribution of *C. punctulata* was similar to the dominant species in that they were captured most often at the borders, yet substantial numbers were collected within field interiors.

Although distances were measured into the fields from east to west, given the shape of the fields and location of transects, beetles might have only dispersed 60-80 m from the windbreak edges. This could have accounted for some of the similarities in beetle catches from 60 to 360 m into the fields. Annual variation in the relative abundance and occurrence of ground beetle species could be expected in both temporary and permanent habitats (den Boer 1986, Luff 1990, 2002). French and Elliott (1999a) showed that annual captures were important separators of ground beetle assemblages, second only to season of occurrence. In this study, the canonical correspondence analysis showed that axes 1 and 2 separated beetle assemblages based on year when they were captured. Axis 1 separated beetle assemblages based on years 2001 and 2002 and axis 2 on year 2000. Beetles were sampled only during August 2000 and probably missed some important autumn breeding species and trap catches such as *H. pensylvanicus*. This was also depicted on axis 1, where *H. pensylvanicus* (Hap) ordinated directly on this axis and toward year 2002. Note also in this study that the two other dominant species (Cya and Ptp) ordinated near the axis origins which indicated the evenness of their relative abundances in all the three years. Ground beetle species closely associated with particular years ordinated near the respective environmental arrow (e.g., Cac and 2001, Chp and 2002). When the effects were factored out of years and fields as co-variables, cropping system was the primary environmental factor separating ground beetle assemblages, as indicated along axis 1.

Other studies had shown cropping systems and management to be important environmental factors affecting ground beetle assemblages (Purvis et al. 2001, Luff 2002). The field border was another important location for capturing ground beetles, as indicated by those species associated with axis 2. Based on partial canonical correspondence analysis, French and Elliott (1999a, b) and French et al. (2001) were also able to classify ground beetles as habitat edge or interior species. For example, in this study, *H. pensylvanicus* ordinated near the axis centres of the partial canonical correspondence analyses, indicating a habitat generalist. Here *H. pensylvanicus* also ordinated near the axis centre, indicating a habitat generalist, but with a slight tendency toward cornfields. Also in this study, ground beetle species closely associated with particular environmental variables (crop type and distance) ordinated near the respective arrow (e.g., Cac and Soybean, and Hae and Do/border). Note also in this study that the three dominant species of ground beetles (Cya, Pol and Ptp) and many of the common species ordinated near the axis origins, indicated the evenness of their relative abundances over both crops and distances. A small number of species accounted for a large portion of all ground beetles captured in all years and crops. *C. alternans, H. pensylvanicus* and *P. permundus* were consistently captured in relatively greater numbers, and along with several other species, were captured throughout the cornfields and soybean fields. Ground beetle assemblages were separated primarily by year and then by crop. The three dominant species and most of the abundant species ordinated near the axis origins, indicating stability over time and ability to occupy multiple habitats. Many species ordinated near field borders and probably represented edge species. It was not clear whether the abundant species spent their entire life cycle within the cornfields and soybean fields or overwinter in the field edges and disperse into the cornfields and soybean fields with time. However, their high relative abundances, continuous seasonal activity, predatory nature, and ability to occupy field centres make these carabid beetles good candidates for biological control of primary and secondary agricultural pests.

2

Economic Importance of Different Species of Ground Beetle

Ground beetles generally benefit agriculture as predators of insects at or just below the soil surface. While most species are opportunistic predators or scavengers, a few species had a restricted food range [e.g., adults and larvae of *Calosoma* fed only on Lepidopterous larvae, and those of *Loricera pilicornis* Fab. on collembola (Thiele 1977)]. Some omnivorous species consumed plant materials as well as insects (Balduf 1935, Lund and Turpin 1977a, Larochelle 1990, Cromar et al. 1999); most of these species belong to the genera Amara, Bonelli and Harpalus, Latreille (Cornic 1973). While adults and larvae of species of these genera often feed on seeds and pollen (Alcock 1976, Best and Beegle 1977, Lund and Turpin 1977b) damage is usually economically insignificant (Thiele 1977). Under stress conditions such as during a drought, ground beetles will feed on fleshy fruits for moisture. Tischler (1958) found that all or almost all of the ground beetles found in agricultural sites in western Europe originated from littoral habitats. Also, despite the inability of most ground beetles in agricultural habitats to climb on plants, ground beetles nevertheless were significant predators of aphids that fall onto or move over the soil surface (Griffiths et al. 1985, Lys 1994).

In eastern Canada, until the late 1960s the use of pesticides was rather uncommon, as the most widely planted crops (e.g., alfalfa and clover) did not require pesticides. However, with the development and spread of new varieties of crops such as corn in the 1970s and soyabean in the late 1990s, at least herbicide applications were commonly applied. In Europe, several studies showed marked decreases in species and abundance of ground beetles over periods of 30 or more years (Kromp 1999). In the prairie region and in south-central British Columbia, wild prairie meadows shared numerous species with agricultural fields. In these regions, the fauna of ground beetles consisted entirely or mainly of native species (Frank 1971, Kirk 1971, Doane 1981). In eastern Canada and coastal regions of British Columbia, the fauna of agricultural ecosystems is quite different from that seen in the prairie regions.

Species introduced unintentionally from Europe dominates in agricultural ecosystems (Rivard 1974, Levesque and Levesque 1994, Raworth et al. 1997).

2.1 Factors Favouring Ground Beetles in Agricultural Ecosystems

Increasing the abundance of predatory ground beetles is likely to benefit agriculture (Thiele 1977). There are two basic approaches to gain this benefit: bioaugmentation by the addition of exogenous beetles, or, enhancement of local populations by provision of suitable habitat. Due to difficulties with rearing, bioaugmentation with ground beetles is not feasible. Larvae are fierce predators, readily cannibalize each other, and must therefore be reared singly. Enhancement of populations by habitat modification have been shown to be effective and may have additional benefits. Lys (1994) clearly demonstrated that weed strips increased the populations of ground beetles in agricultural settings. The strips increased the variety and quantity of prey, and more importantly act as a refuge against high temperatures and low moisture. Lys (1994) also found highest ground beetle densities to be 12 metres.

For continuous enhancement of ground beetle populations over the years, tillage practice, cropping sequence, habitats around field edges, and the type of crop planted are significant. In southern Ontario, Tyler and Ellis (1979) studied ground beetles in fields with no tillage, minimum tillage and conventional tillage. Only conventional tillage reduced populations of about half of the most common species of ground beetles. Carcamo et al. (1995) found no differences in abundance of captures, but Tyler and Ellis (1979) found that some species were affected. Holliday and Hagley (1984) made similar observations in Ontario apple orchards. In Canada field edges were either quite narrow between cultivated fields, or much wider if the field was located next to either a forested area, a meadow or a fallow field. Well-drained field edges can offer excellent wintering sites for ground beetles (Booij 1994). In Europe (Thiele 1977) and in North America (Doane 1981), drained grassy field edges had similar ground beetle fauna to that of typical agricultural sites and, therefore, could serve as sources for ground beetles associated with agriculture. Grassy field edges were important refugia for ground beetles, parasitoids and other predatory insects, and should not be sprayed with pesticides if pest control proved necessary in the cropped field. Cropping practices affect ground beetles where the fauna of perennial crops were similar in composition to that of annual crops, but the arrangement of dominance among species was completely different (Thiele 1977). Annual crops that rapidly shaded the exposed soil with a dense canopy of leaves attenuate extremes of temperature and moisture and had the highest population of ground beetles (Varis et al. 1984).

Effect of Pesticides on Population Buildup of Ground Beetle

Pesticides (insecticides, herbicides, fungicides, nematocides) are applied to many Canadian agricultural crops. While many crops were not sprayed, some required one insecticide application, and a few required several pesticides applications [e.g., in addition to herbicides and fungicides, up to six insecticidal sprays are applied each year on apples (information gathered from apple growers at the Agriculture and AgriFood Canada Research Station near Frelighsburg, Quebec)]. Ground beetle populations recover relatively rapidly after the application of non-persistent insecticides, but only very slowly following persistent insecticides (Thiele 1977). Though less toxic than insecticides, herbicides and nematocides can reduce the population of many species of ground beetles (Thiele 1977, Freitag 1979, Gregoire-Wibo 1983). At the time of reproduction, ground beetles are very active and cover marked distances each day (Rivard 1966). Therefore, a few days after a pesticide spray, there is a mixture of ground beetles consisting of specimens exposed to the spray and newly migrated specimens from marginal sites not exposed to the spray. The distance of sprayed fields from a source area will influence the apparent impact and the recovery time. As reported for other species in Europe (Heydemann and Meyer 1983), the populations of spring breeding ground beetles were most affected as the pesticide application were mainly in the spring. Sublethal effects of pesticides on fertility and fecundity of ground beetles warrant further study.

2.2 Role of Ground Beetles in Pest Management

Ground beetles can play a significant role in ecologically-based integrated pest management programmes that focus on avoidance or reduction of pest pressure through cultural practices and biological controls. Carabids are common in agricultural fields in the Northern Hemisphere. Since an early report by Forbes (Forbes 1883), they have generally been considered beneficial natural enemies of agricultural pests, although a few species are pests themselves (Luff 1987, Thiele 1977). These insects are considered to be mostly opportunistic feeders that consume a variety of foods; however, the majority of species were as primarily predatory, feeding on other insects and related organisms. Most species located food by random search, although some day-active (diurnal) species hunted by sight. Larvae and adults typically have similar feeding habits; however, larval diets are more restricted due to a limited search range underground. The natural diets of carabid beetles are still widely undetermined. Laboratory studies have shown that carabid beetles will eat nearly anything offered, however they typically show food preference and it is unclear whether or not these feeding habits are typical in nature (Tooley and Brust 2002).

Prey can change throughout their life cycle based on nutritional needs or a change in the resources or environment. Several ground beetle species are phytophagous (feed on plants). Of particular interest is "seed predation," where plant seeds are not only consumed by ground beetles, but destroyed in the process (as opposed to merely ingesting the seed). It has been suggested that plant feeding (herbivory) and weed seed predation (granivory) is largely underestimated in ground beetles (Tooley and Brust 2002). Weed seed predation by beetles can potentially be used to lower costs associated with weed populations and to increase crop yield. A number of carabid species were identified as weed seed predators. They were shown to consume a variety of agriculturally important weed species seeds including common ragweed, common lambs quarters, and giant foxtail (Lundgren 2005). Seed preferences may be due to the differential oil content of seeds as well as the ability of the beetles to handle the seeds. Post-dispersal consumption of seeds (feeding on fallen seeds) is more common among ground beetles in comparison to pre-dispersal consumption. Overall, weed seed predators may alter the plant species composition in an area rather than eliminate a weed species.

Changing the species composition can give crop species a better chance of success by reducing competition for resources such as light, nutrients and water, thus increasing crop yields. Understanding weed seed preferences is critical to effectively utilizing weed seed predators in biological control programmes. Gut analysis of ground beetles for different weed seed proteins and laboratory food preference studies may help reveal these relationships. Weed seed predation by arthropods such as ground beetles can potentially be used to lower costs associated with weed populations and to increase crop yield. The ground beetles are of great importance in the bio regulation of insect pests, but their significance are still not assessed precisely. Most members of the family Carabidae are primarily carnivorous; larvae as well as adults are nocturnal and, hence, less well known. Prior to taking up steps to conserve and augment these bio-agents in agroecosystems, extensive surveys to evaluate the diversity and abundance of predatory carabids in different crop ecosystems shall become necessary to identify the dominant species that can be conserved for future. The survey shall also enable one to know about the resident species. Studies on their biology must also be taken up for proper utilization of the species concerned as a bioagents. Low-input farming with reduced tillage or otherwise biologically managing the farms shall become a prerequisite to enhance the predatory activity of these predatory beetles. Diversified cropping followed with good ground covered by harbs is preferred by this carabid beetle. Use of synthetic pesticides have to be avoided to safeguard these natural enemies.

The soil hydrological regime, soil treatment and crop cultivation determine the carabid population structure. The dominance structure and seasonal population dynamics of carabids varies according to the crop type and density or ground cover. The exploration of conditions under which generalist predators can limit prey revealed that such predators are self-damping and highly polyphagous and that their lifecycle are not in synchrony with their prey. They could suppress pest outbreaks, but in general, their major beneficial role is to prolong the period between pest outbreaks i.e. when the pest abundance was in the so-called natural enemy ravine (Southwood & Comins 1976). To increase carabid's effectiveness, biological control practitioners shall consider the general habitat favourability that will keep them near their required site of action. The effectiveness of a natural enemy can be established through several sequential steps (Luck et al. 1988, Sunderland 2002, Wratten 1987): (1) evaluating dynamics and correlating predators and prey density, (2) obtaining direct evidence of a tropics link between the prey and the predator, (3) experimentally manipulating the predator density and its effect on pest numbers and (4) integrating the above information to quantify the effect of predator on prey.

Factors Affecting Predation of Ground Beetles

Seasonal rhythms involving dormant periods during winter and/or summer (aestivation) are an integral part of the life history of temperate-region ground beetles. The activity of the two most typical groups peak in either spring or autumn. The peak usually coincide with the reproductive period, although the connection between activity and reproductive rhythms is flexible in many species (Makarov 1994). Such rhythms are inseparable from individual, especially larval development. The study of searching behaviour showed that many carabids presumably find their food via random search, several diurnal species hunted by sight (Parmann 1986). Other species use chemical cues from aphids (Chiverton 1988), springtails (de Ruiter et al. 1989) or snails (Wheater 1989) to find prey. Consequently Tiger beetles always keep its antennae in the same fixed position: straight ahead, angle at a V, and held straight above the ground. The antennae can move, but they never do while the beetles in motion. Actually the antennae are obstacle-detectors. If they hit an obstacle, their flexible tips bend back before springing forward again. The beetle move too fast to change course, but it can tip the body slightly upwards so that it skitters over the obstacle rather than running head long into it. Because of their shape, the antennae can slip over the edge of an obstacle, which tell the beetles that there's a top they can run over. If their antennae are intact, they clear the obstacle most of the time, even when the insect is painted over their eyes.

The insect is specialized for many modes of life in and on the ground, as well as on plants or as "miners" spent most of their time digging underground. Agricultural setting tend to support ground beetles that are likely to feed on insect pests such as caterpillar and aphid, as well as weed seeds. Farmers have long been encouraged to look upon this carabid beetles as naturally occurring pest control. Although it is difficult to say exactly how much pest control is provided by these beetles on individual farms, it is safe to say that they have a meaningful impact like other general predators such as ladybeetles and spiders. The beetles play a major role in agro ecosystems by contributing to the mortality of weed seeds, insects, and slugs. They can consume up to their body weight daily. They eat a wide variety of pest organisms including aphids, moth larvae (such as armyworm, cutworm and gypsy moth larvae), beetle larvae (such as the corn rootworm, colorado potato beetle and the cucumber beetle), mites, and springtails. It has become increasingly clear that ground beetles are important polyphagous natural enemies in agricultural landscapes that have the potential to suppress major insect pest species from reaching outbreak levels. Moreover, if augmented through their conservation, they will restrict minor insect pest species from becoming major ones. Many studies have analyzed the importance of habitat characteristics, management practices such as pesticide applications and tillage frequently which reduce the abundance of this insect. Organic and low-input production systems usually sustained more abundant beetle communities than conventional system.

Conserving ground beetles through habitat manipulations and cultural practices can enhance the natural regulation of arthropod pest and weed populations, reducing the need for chemical controls. The key to take the advantage of the benefits from this carabid beetles in agriculture is improving their survival capacity. Certain farming practices can help to conserve carabid beetle populations. These insects are more likely to survive in fields where non-inversion (e.g., chisel plough) tillage is used. In comparison with inversion tillage practices (e.g., moldboard plow), non-inversion tillage cause less soil disturbance and thus, less direct mortality of the eggs, larvae and adults. Due to their relatively soft bodies, carabid beetle larvae are especially sensitive to tillage practices. Minimum tillage systems also preserve surface vegetation and mulch, which can provide microhabitats for beetles as well as protection from environmental conditions and other predators. Both chisel plowed and no-tillage systems show approximately double the activity of weed seed predators than conventional tillage systems (Shearin et al. 2007). The choice of mowing technique is also important for conserving the beetles. Flail mowers at a 5 cm cutting height are shown to reduce numbers of ground-dwelling arthropods by 50% whereas bar and rotary mowing do not cause significant damage (Humbert et al. 2008). Habitat management also play a critical role in conserving ground beetles (Menalled et al. 2001). For ground beetles to survive and reproduce, they need a protected place to overwinter, mate, and lay eggs. That habitat shall also provide food (arthropods, mollusks and plant seeds), a favourable microclimate, and shelter from other predators. While cover cropped fields can provide excellent winter cover for ground beetles, the eventual mowing or tillage that occurred may make these environments unsuitable for long-term conservation.

Ground beetle habitat is a permanent piece of vegetated land. Field edges, marginal lands or select areas within a crop field are excellent sites for ground beetle conservation. Creating a "beetle bank" involves raising a 2-5 ft wide bed, seeding the bed with a native perennial grass mixture (approximately 30 lbs/acre broadcast seeded), and laying down a mulch layer (Ellen undated). Carabid beetles were shown to be twice as abundant in crop fields adjacent to beetle banks and other uncultivated habitat than in fields without this habitat (Hance 2002). Generally, stands of diverse perennial plants, which offer a diversity of microhabitats and food resources, will support diverse carabid beetle communities. Studies had also shown that utilizing different species at different times of the year could improve biological control (Tooley and Brust 2002). Proper organic fertilization and green manure application enhance carabid recruitment. Intensive nitrogen amendment may indirectly affect carabits by altering crop density and microclimate. They are enhanced by crop diversification in term of monocrop heterogeneity and weediness as well as by intercropping and the presence of field boundaries or farm scraping. The major factors conditioning the association of ground beetles in agroecosystem are: (a) The temporal stability of the habitat with stable habitats providing suitable and sustainable environment. (b) The reduced tillage which enhances carabid diversity and abundance. Organic or biologically managed farms are more suited for conservation (c) Early crops and crops with greater cover favour carabid beetle abundance. (d) Crop diversification through intercropping/ multiple cropping, farm scraping and mulching positively affect the population build up. (e) Beetle activity is known to be correlated with hunger levels and availability of preferred prey. (f) The soil hydrological regime, atmospheric humidity and temperature affect carabid diversity and numerical abundance. Sharma and Bisen (2013) reported *Chlaenius* sp. as predator of *Laphgma pyrausta nubilalis* in vegetable ecosystems in Madhya Pradesh, India.

2.3 Ground Beetles as Common Insect Predators

Carabid beetles can have beneficial impacts on agriculture. By consuming a variety of insect pests they can help to protect crops from pest damage and associated losses, and decrease costs associated with pest controls. The types of carabid beetles found at a location can also be a valuable biological indicator to assess the impacts of different habitat management and tillage practices. To capitalize on the potential benefits from this type of predator, land managers can conserve habitats that are beneficial to the survival of the beetles. Informed decisions regarding tillage practices, pesticide use, and the establishment and management of beneficial habitats can enhance carabid beetle abundance and diversity (Dreisig 1981). Dissection of several thousand individuals of 24 European species (Hengeveld 1980b) revealed the remains of aphids, lepidopteran larva and adults, fly larvae, heteropteran, opilionids, beetles and springtails. The per cent increase in the body weight of beetles fed with aphids (24.1%) was significantly higher than that in the beetles fed with termites (14.4%) and *Corcyra* larva (11.7%). All species in Hengeveld's study (1980b) were polyphagous and consumed plant materials in addition to other food items.

2.3.1 Genus: CHLAENIUS

Chlaenius velutinus **(Duftschmid)**

Body 13 to 15 mm (0.59 inches) in length, metallic green, covered in fine hair, like velvet on the elytra, head 0.69 and 0.70 width of prothorax, narrower than thorax, mouthpart biting and chewing type, mandibles short, subtransverse, front closely punctuate with mixed moderately coarse and fine punctures, prothorax subquadrate, more narrowed in front than behind; width/length 1.39 and 1.34; base/apex 1.33 and 1.29; side margins very narrow, each with seta at basal angle, disc with impressed middle line and slight sublinear baso-lateral impressions, surface of disc coarsely slightly irregularly punctuate (Fig. 14). Width elytra/prothorax 1.34 and 1.37, striae impressed slightly punctulate, intervals flat or slightly convex, coarsely punctuate, punctuation of inner intervals anteriorly especially coarse and irregular with only a single irregular line of punctures in places of anterior half of each of first four intervals. Structure of pronotum varied among the species as given in Fig. 15. Innerwing fully developed. Lower surface extensively but irregularly punctuate, leg long and slender. The beetle tend to be found on or near the ground and emits foul-smelling when handled.

The insect passes through four separate stages of growth: egg, larva, pupa and adult. On an average, the beetles produce one generation per year. In most ground beetle species, females lay eggs in soil. After finding a suitable site, females will singly deposit between 30 and 600 oval shaped eggs within the soil or in the layer of plant residues on the soil surface. Protected egg sites are very important because young larvae have limited mobility for finding food and their relatively soft bodies are vulnerable to predators. Parental care, including egg guarding have been observed in female that produced small

Fig. 14. *Chlaenius velutinus* (Dufts.) (Image credit: Author) (See Plate 1)

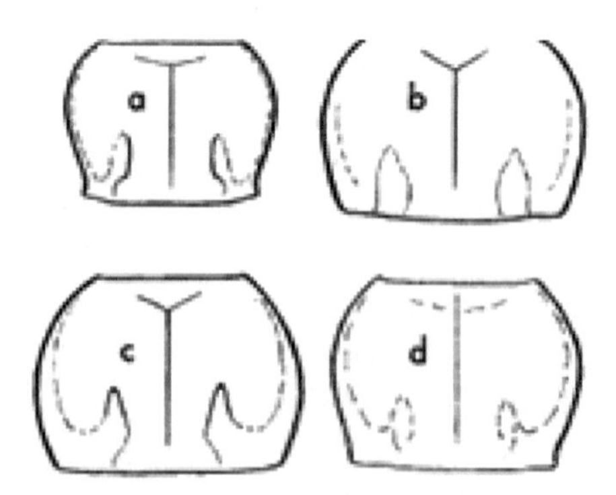

Fig. 15. Pronotum of (a) *Chlaenius vestitus* (b) *C. tristis* (c) *C. nigricornis* and (d) *C. nitidulus* (After Carl H. Lindroth, *RES*1974)

litters. Upon hatching, larvae feed and grew for 1-2 years and pupate in small chambers made of soil. Species are sometimes distinguished as either having winter or summer larvae. Larvae live entirely under the soil surface, where they pupate usually after three larval stages. Adults can live between one and four years. Larger species, as well as those that overwinter as larvae, tend to have the longest life spans (Lovei and Sunderland 1996). The beetles have functional wings, flight is used primarily for dispersal, such that they spend nearly their entire lives on the ground. Adults are fast-moving predators that feed on insect pests of cereal and vegetable crops. Adults were also observed climbing plants in search of prey. The larvae are also predators, and most species hunt in the same way as the adults, by patrolling at night and hiding during the day, although many ground beetle larvae tend to remain undercover even while hunting: some hunt under fallen leaves, others hunt underground.

Some other species viz. *Chlaenius micans* (F.) and *Chlaenius posticalis* Motschulsky were also recorded as the predator of diamond-back moth larvae in Japan (Suenaga and Hamamura 1998) and the study showed that the female adults of *Chlaenius micans* (F.) and *Chlaenius posticalis* Motschulsky consumed 191 and 92 early fourth instar larvae of diamond-back moth respectively during entire larval period (Table 7).

Table 7. Consumptions of early 4th instars of diamond-back moth (DBM) by *Chlaenius micans* (F.) and *Chlaenius posticalis* Motschulsky and duration period of carabid larvae (mean ± SE)

Carabid instar	*No. of DBM larvae consumed*		*Larval period*	
	C. micans	*C. posticalis*	*C. micans*	*C. posticalis*
1st	7.0±0.9	4.0	3.5±0	3.5
2nd	20.8±2.6	11.7±1.6	2.8±0.3	3.0±0
3rd	163.7±17.0	75.2±8.9	12.2±0.6	9.8±0.6
Total	190.8± 8.3	87.6±8.3	18.3±0.6	12.8±0.6

Source: Suenaga and Hamamura 1998

Both the predators were sometimes observed crawling on the cabbage leaves in daylight or night hours but were not observed feeding on the diamond-back moth larvae. In general the *Chlaenius* sp. showed less effectiveness in killing of larvae of codling moth *Cydia pomonella* (L.) in laboratory test (Riddick and Mills 1994) whereas *Chlaenius erythropus* Germar exhibited low percentages in positive reactions to antiserum prepared from corn earworm *Heliothis zea* (Boddle) (Lesiewiez et al. 1982). Adults of these species might be rather effective in suppressing soil hiding noctuid larvae such as cabbage armyworm *Mamestra brassicae* (L.) and the common cutworm *Spodoptera litura* (F.) in Japan (Suenaga and Hamamura 1998). The insect is swift runner and moves rapidly over the plant and soil surface, especially at night, in search of suitable prey. Often these predatory beetles attack prey larger than themselves. It is recorded as common predator of *Spodoptera exigua* (Shephard et al. 1999).

Although adult ground beetles may be fast enough to escape from large disturbances like heavy plowing, their larvae and eggs are not mobile and are very susceptible to these disturbances. Increasing amount of ground cover in crop fields benefitted the ground beetles and cover crop enhanced beetle numbers. However, cover crops eventually turn under and destroy and the adults need refuges where they can thrive and reproduce year round. Good refuges for ground beetles can be created in the borders of the field by planting a diverse mix of annuals and perennials (Mennaled et al. 2001). A similar strategy is to create these "beetle banks" within the crop fields itself. Other strategies, such as intercropping and using organic fertilizers, seemed to promote ground beetles as well (Kromp 1999). Among the different species *Chlaenius chlorodius* Dejean is common in India (Saha et al. 1992) which was recorded as a predator complex of the teak defoliator, *Hyblaea puera* Cramer (Lepidoptera: Hyblaeidae) at Veeravanallur, Tamil Nadu, India (Loganathan and David 1999).

Chlaenius rufifemoratus MacLeay

Adults are black, medium sized, more or less pubescence on dorsal and ventral surface, more or less truncate apical segments of palpi, dense pubescence of antennae from 4th segment, labrum with six setae, clypeus with two setae, pronotum with a pair of posterior fined setae, elytral base wider than that of pronotum, striae eight complete and mentum with median tooth. Head green with coppery reflections, elytra black with two yellow patch one each on anterior and posterior region. Male can be readily distinguished from the female by the three dilated joints of the pro-tarsi (Fig. 16).

Fig. 16. *Chlaenius rufifemoratus* MacLeay (Image credit: Author) (See Plate 1)

The adult is an important biological control agent in agro ecosystems. With their large eyes, spiny powerful legs, and large jaws, they are formidable predators in the insect world. They live on the surface of the soil where they capture and consume a wide assortment of soil dwelling insects, including caterpillars, wireworms, maggots, ants, aphids and slugs. As early as 1883, Forbes reported aphids to be component of this carabid diet. In addition to this the beetle also ate the seeds of troublesome weeds and are considered one of the "many little hammers" that help regulate weed populations (Liebman and Gallandt 1997). The adult also consume Japanese beetle (E); black cutworm (L); European corn borer (L); common stalk borer (L); armyworm (L); fall armyworm (L) as available from http://ento.psu.edu /extension /factsheets/ground-beetles.

Larvae of *Chlaenius viridis* Men. were observed to prey upon the soybean leaf webber *Lamprosema* sp. in the field while under laboratory conditions, a single adult consume nine tobacco caterpillar (*Spodoptera litura* (F.) larvae per day. It also equally preys on the larvae of cotton leaf roller (*Sylepta derogata*) (Swaminathan et al. 2001). *Chlaenius panagaeoides* (Laferte) migrate from forests to fields, they were seen to climb plants and eat *Aphis craccivora* in cowpea fields (Rajagopal and Kumar 1992). The insect is a predator of the teak defoliator, *Hyblaea puera* Cramer (Lepidoptera: Hyblaeidae) at Veeravanallur, Tamil Nadu, India (Loganathan and David 1999).

Chlaenius virgulifer (Licini)

The beetles are medium to large size, mostly brilliant colour, more or less pubescence on dorsal and ventral surface, more or less truncate apical segments of palpi, dense pubescence of antennae from 4th segment, labrum with six setae, clypeus with two setae, pronotum with a pair of posterior fine setae, elytral base wider than that of pronotum and striae eight complete and mentum with median tooth (Fig. 17).

Fig. 17. *Chlaenius virgulifer* (Licini) (Image credit: Author) (See Plate 1)

The insect is a fast-moving predator that feed on small insects, spiders and other arthropods. They usually hunt at night by patrolling the ground, and are found in a variety of habitats, including farmland, wooded areas, and lawns. During the day, the adults hide under rocks, logs, and fallen leaves. Rao (1971) recorded that this insect was reported as an important predator of the larvae of *Chilo partellus* (Swin.) in South East Asia. Wingo et al. (1975) reported that abundance and time of initial appearance of immature stages of face fly *Musca autumnalis* De Geer at pasture manure depended upon the predation of *Chlaenius tomentosus* Say. Rajagopal and Kumar (1988) had studied the predation potential of *Chlaenius panagaeoides* (Laferte) on cowpea aphid in India. Shanower and Ranga (1990) reported that Larvae of *Chlaenius* sp. were found to attack and ate larvae of gelechiid *Aproaerema modicella* inside mine during the rainy season of 1988 at Patancheru, Andhra Pradesh, India, in groundnut and soyabean leaves. *Chlaenius emarginatus* Say was observed to attack the gypsy moth *Lymantria dispar* (L.) (Lepidoptera: Lymantriidae) in Pennsylvania, USA (Maclean and Usis, 1992). However, as generalist natural enemies, they may be better suited for prolonging the period between pest outbreaks than for quickly reducing a pest population whose density have already exceeded an economic threshold. Collectively, generalist predators, like carabid beetles, can prevent damage to crops by as much as 40%, compared to areas where generalist predator numbers were kept experimentally low (Clark et al. 1994).

Among the different species *Chlaenius bioculatus* Chaudoir, *C. pictus* Chaudoir, *C. leucops* (Wiedermann) (Park et al. 2006), *C. hamifer* Chaud., *C. circumdatus* Brulle, *C. chlorodius* Dejean and *C. laevipennis* Chaudare (Saha et al. 1992) were common in India and *C. tomentosus* Say, *C. pusillus* Say, *C. platyderus* Choudoir, *C. lithophilus* Say and *C. nemoralis* Say were recorded from USA (Esau 1968). *Chlaenius tomentosus* Say would thus appear to fare best in grassy habitat which might be somewhat disturbed. The range was the eastern United States, east of the Rocky Mountains and also Arizona and New Mexico. Bell (1960) indicated that only one specimen had been recorded in a pine forest, except for the mass nocturnal flights, other having been found beneath stones in several different situations, usually disturbed by man. The author also found this insect under vegetation in dry, grassy meadows. Rivard (1964a) recorded this insect from dry, open ground with moderate to dense vegetation. The modified mouthparts of *C. pusillus* Say suggest that they must be active predators of large soft-bodied prey.

The range is the south-eastern United States and adjacent northern areas, Connecticut, New York State, Michigan and Iowa. The southern limit is from Florida, and throughout the gulf States to Texas. The habitat adaptability of *C. platyderus* Choudoir was broad in USA where Massachusetts, Minnesota, North Dakota and Colorado represented the northern range limit and Gulf states, north Carolina, west through Tennessee, Louisiana, Texas, and New Mexico in south (Bell, 1960). *Chlaenius lithophilus* Say is known to occur in most of the United States. The northern range is Newfoundland, Quebec, the New England States, New York State, Pennsylvania, Michigan, Minnesota, south Dakota, Colorado, Idaho and Montana, southern extremes were Georgia, Alabama and Texas. The species had been taken by Bell (1960) in a cypress swamp and also in an Illinois rumex swamp. Lindroth (1955) found the species around borders of rivers, oxbow lakes and small pools in Newfoundland. Rivard (1964a) found the species on moist soil with moderate to dense vegetations. Bell (1960) stated that *Chlaenius nemoralis* Say were found and collected under trash, close to river banks and was also quite common beneath boards and stones lying about within the city of Urbana, Illinois, which was far from any open water body. The range is the eastern United States, Minnesota, down to western Kansas, to central Texas; southern limits are Florida and Alabama.

2.3.2 Genus: BRADYBAENUS

***Bradybaenus* sp.**

The genus Bradybaenus is characterized by: head with frontal furrows represented by a small pit (without clypeo-ocular grooves), mentum without median tooth, paraglossae pubescent, penultimate articles of labial palpi each with three or more setae on anterior margin, mesotarsomeres of male not dilated, without rows of scale-like bristles ventrally, metatarsomere 1 longer than 2 but markedly shorter than 2 and 3 together, anterior angles of pronotum without setae, elytra with scutellar stria on interval 2, abdomen with penultimate and two preceding sterna with a row of long setae near hind margin, median lobe of aedeagus with ostium not deflected to left (Facchini & Sciaky 2004).

Bradybaenus sp. is one of the predator of termite in India (Kumar & Rajagopal 1997). Among the different species *Bradybaenus perrieri* Jeannel and *Bradybaenus opulentus* Boheman are predominant in different crop fields (Figs 18 and 19) of south east Asia.

Fig. 18. *Bradybaenus perrieri* Jeannel (From https:// commons. wikimedia.org)

Fig. 19. *Bradybaenus opulentus* Boheman (From https://commons. wikimedia.org)

2.3.3 Genus: CRASPEDOPHORUS

***Craspedophorus angulatus* (Fab.)**

Häckel & Kirschenhofer (2014b) treated *Craspedophorus* as the genus subdivided into 13 species groups in the Oriental fauna, all defined mainly based on colouration, size and shape of the body, pronotum, elytra and elytral maculae. Also, many Oriental species like *C. angulatus* (Fab.) share a conspicuous microsculpture of the forebody dorsum including the labrum and the exposed parts of closed mandibles, the neck being either rugose or coarsely punctuate and the penultimate labial palpomere subtriangular and plurisetose at inner margin, i.e., with three or more setae arranged in a subtransverse row.

Craspedophorus angulatus (Fab.) is a predator (Fig. 20) of the teak defoliator, *Hyblaea puera* Cramer (Lepidoptera: Hyblaeidae) in an intensively managed teak plantation at Veeravanallur, Tamil Nadu India. The pest can reduce teak yield by 44% in South India (Loganathan and David 1999).

Fig. 20. *Craspedophorus angulatus* (Fab.) (From https:// commons.wikimedia.org)

2.3.4 Genus: AMARA

Amara sp.

Body length of male 6.8–7.0 mm, female 6.6–7.0 mm. Colour of body dark brown, antennae, palpi and legs reddish brown. Dorsal microsculpture comprise isodiametric or nearly isodiametric sculpticells throughout, very faintly impressed on head in both sexes, more shallowly impressed on pronotum and elytra in males than in females; males with shinier dorsal lustre than females. Head smooth, broad, with distinct, hemispheric eyes, pronotum slightly transverse, with the greatest width slightly anterior to middle and posterior margin narrower than the base of elytra; lateral margins more rounded on anterior half, less arcuate or nearly straight in basal half; posterior margin slightly concave in middle; posterior angles distinct, slightly obtuse, narrowly rounded apically; anterior angles rounded, only slightly extended (about the diameter of the second antennomere) anteriorly beyond the front margin; inner basal foveae formed as short, deeply impressed longitudinal grooves; outer basal foveae absent; basal region with scattered, very fine punctures in and around inner basal foveae.

Prosternum of male without a punctate fovea at middle; prosternal intercoxal process smoothly rounded apically, unmargined, asetose apically. Pterothorax with metepisterna short, not longer than width across anterior margin. Elytra with slightly curved sides and finely punctate striae; parascutellar striae short, located between striae 1 and 2 and extended from basal margin near base of stria 2 apicomedially toward stria 1; basal borders nearly straight, very slightly arched forward laterally; humeral teeth small but distinct and sharp; umbilicate setal series sparsely and unevenly spaced in the middle region; stria 7 without subapical setiferous pore punctures. Hind wings short, reduced to a minute scale, hence adults flightless. Legs with all femora bisetose; mesotibiae of males with a well-developed subapical medial tooth; metatibiae of males with a brush-like patch of setae ventrally in apical one-fourth. Abdomen with venter only punctate laterally on the sternites 2 and 3. Male with one pair and female with two pairs of anal setiferous pore punctures at the apical margin of the last visible sternite (Hieke et al. 2012). *Amara* sp is a predators of *Phyllophaga anxia* (LeConte) (Coleoptera, Scarabaeidae) in Quebec, Canada (Poprawski 1994). Among the other species *Amara familiaris, Amara aenea* and *Amara plebeja* are polyphagous predators in Belgium (Loreau1984) whereas *A. torrida* Paykill, *A. pennsylvanica* Hayward, *A. carinata* Le Conte, *A. exarata* Dejean, *A. latior* Kirby, *A. apricaria* Paykill, *A. avida* Say, *A. obesa* Say, *A. ellipsis* Casey, *A. aeneopolita* Casey, *A. littoralis* Monnerheim, *A. cupreolata* Putzeys and *A. angustata* Say were predominant in USA (Esau 1968). The distribution of this insect species seems to be principally in disturbed site. Lindroth (1955) stated that the beetle was confined to open, dry grasslands with a firm soil base.

Fig. 21. .*Amara (Zezea) plebeja* (Gyllenhal) From https:// en.wikipedia.org

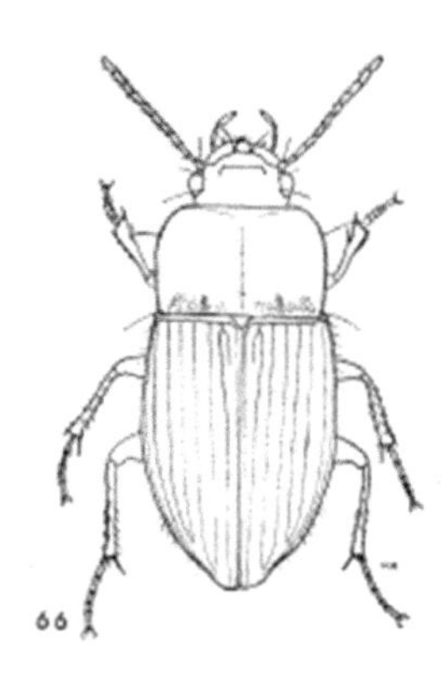

Fig. 22. .*Amara (Zezea) plebeja* (Gyllenhal) Courtesy: Carl H. Lindroth, *RES* 1974

A. exarata Dejean had been recorded from Kansas, Iowa, and east to north Carolina, and long Island in New York State and north to Massachsetts (Esau 1968). The species was collected only in disturbed areas; the fencerow and cornfield but primarily in the fencerow. Everly (1938) had collected the species in an Chio sweetcorn field. Dambach (1948) reported it from bluegrass sod fencerows. Lang (2003) reported that *Amara (Zezea) plebeja* (Gyllenhal), a common predator (Figs 21 and 22) of wheat aphid like *Sitobion avenae* (F.), *Metopolophium dirhodum* (Walk.) and *Rhopalosiphum padi* L. in Munich, Germany. *A. latior* Kirby was reported from Newfoundland, Nova Scotia, south as far as Texas, west to California, Oregon, and British Columbia (Lindroth 1955, Hayward 1908). *A. apricaria*, *A. fulva*, *A. equeatris*, *A. quenseli*, *A. bifrons*, *A. lunicollis*, *A. aenea*, *A. ovata*, *A. montivoga*, *A. nitida*, *A. communis*, *A. familiaris*, *A. lucida*, *A. plebeja*, *A. similata* and *A. infima* (Figs 23, 24 and 25) were in the record of Lindroth (1974). The insect prefers open ground in general with dry soil.

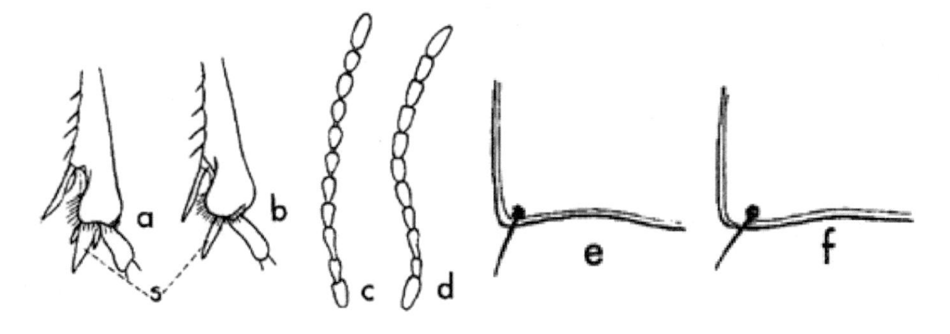

Fig. 23. *Amara*. Apex of front-tibia in (a) *plebeja*, (b) *similata* (s, terminal spur). Antenna of (c) *infima*, (d) *tibialis*. Hind-angle of pronotum in (e) *ovata* and (f) *nitida* (After Carl H. Lindroth, RES 1974)

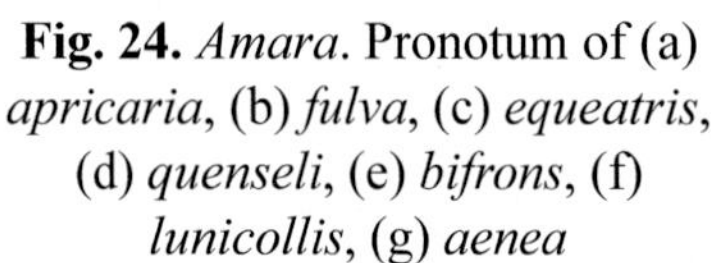

Fig. 24. *Amara*. Pronotum of (a) *apricaria*, (b) *fulva*, (c) *equeatris*, (d) *quenseli*, (e) *bifrons*, (f) *lunicollis*, (g) *aenea*

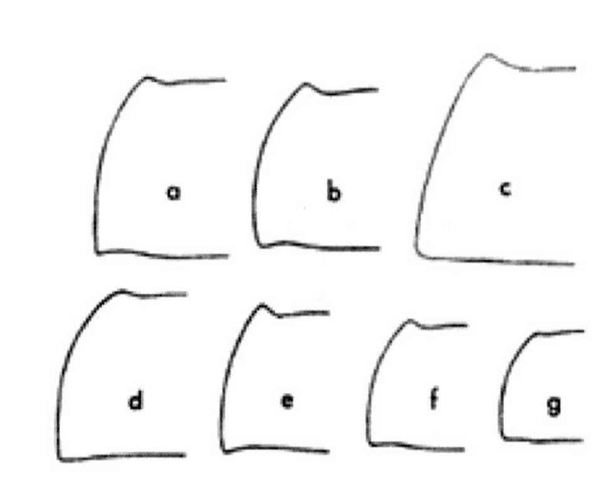

Fig. 25. *Amara*. Pronotum of (a) *similata*, (b) *ovate*, (c) *montivoga*, (d) *nitida*, (e) *communis* (typical), (f) *familiaris* and (g) *lucida* (After Carl H. Lindroth, *RES* 1974)

Rivard (1966) found it common in agricultural land; Lindroth (1955) reported that the species had been taken on dry grassland with gravelly soil. *A. apricaria* was found throughout Eurasia and was in north America as an introduction. On this continent it was known from Nova Scotia to British Columbia (Lindroth 1955). The insects were found under debris of a sweet corn field in Ohio (Everly 1938) and under stones in potato field (Procter 1946). *A. avida* Say was known from Nova Scotia to Edmonton, Alberta and British Columbia. The southern limit of the range seemed to be Virginia, Indiana, Illinois, Iowa, Colorado and British Columbia. Lindroth (1955) reported that this species chooses intimate association with human culture in Newfoundland, being found in areas of scarce vegetation and especially weedy places. The species was most commonly found on open ground with dense vegetation (Rivard 1964a). This insect was found under stones of a potato field in the Mt. Desert area of Maine (Procter 1946). *A. obesa* Say was trans-American in a rather broad belt from Newfoundland and Nova Scotia, west to British Columbia, south to the western part of Oregon, Idaho, Arkansas and New York States.

Walkden and Wilbur (1944) collected specimens from a river bottom, sweet-clover field, alfalfa field, bluegrass and roadside bromegrass pasture. Lindroth (1995) collected some specimens on a sandy seashore. Rivart (1964a) found the species on open ground with dense or moderate vegetation. Parker and Wakeland (1957) reported *Amara obesa* adults to had been reared from larvae feeding on grasshopper egg pods collected in Klamath Country, Oregon. Since *A. obese* was widespread, Parker and Wakeland felt that it might be the most important carabid predator of grasshopper egg pods. *A. ellipsis* Casey was recorded from Iowa, Kansas and Colorado (Casey, 1918b). The disturbances caused by surrounding agricultural practices favoured the population development at other places. Lindroth (1955) stated that *A. aeneopolita* Casey was widely distributed in northern latitudes, but is known only from Newfoundland, Labrador, Quebec, Michigan and Manitoba. *A. littoralis* Mannerheim was distributed in Alaska (Hayward 1908), Iowa, Curtiss farm fencerow, Dekalb, Kalsow and Cayler Prairie (Esau 1968) of USA. Vestal (1913) stated that the species fed mostly on vegetable matter. *A. angustata* Say was known from Canada, the New England States, south to Virginia and Indiana and West to Kansas (Esau 1968).

2.3.5 Genus: CARABUS

Carabus olympiae Sella

The body rather long and flattened, with a clearly differentiated head, thorax and abdomen, long, slender legs, and quite prominent mandibles and palps. Although not as spectacularly coloured as some of its relatives, *Carabus olympiae* Sella is still an attractive insect, with a dark purplish-blue head and pronotum, the margins of which have a golden sheen, and brilliant, metallic, golden-green elytra, with coppery or purple margins whereas *Carabus hispanus* Fabis dark black (Fig. 26). Most active at night particularly during periods of thunderstorms and high humidity. Believed to have originally been a forest-dweller, *Carabus olympiae* Sella occurs in more open habitats, ranging from beech forest to open shrubland, at elevations of around 800 to 2,000 metres. It may have a preference for beech forest as well as areas of alpen rose (*Rhododendron ferrugineum*) and bilberry (*Vaccinium myrtillus*) but avoided pastures. The main period of activity of *Carabus olympiae* Sella ran from June to September, peaking in July (5), with the eggs laid in late spring or early summer.

Fig. 26. Carabus hispanus Fab. (From http://www.wikiwand.com)

The number of eggs is relatively low, each female lays a maximum of around 29. The larvae of *Carabus olympiae* Sella pass through three developmental stages before pupating, with total development taking approximately one to one and a half months. The young adults emerge at the end of the summer and feed for about a month, built up fat reserves before entering diapause during the winter. Eggs lay late in the year may pass the winter as larvae, and it is possible that some individuals of this species may delay reproduction if conditions are unfavourable, or even reproduce in more than one successive year (http//www.carabidae.ru). *Carabus olympiae* Sella is a voracious predator. Both adults and larvae actively hunt for invertebrate prey, and are thought to have a preference for molluscs, in particular the snail *Arianta arbustorum*. Consequently *Carabus granulatus* Linn. depressed aphid number in winter wheat field in Munich Germany (Lang 2003).

Carabus nemoralis Mueller

Carabus nemoralis (commonly called the "Bronze Carabid") is a ground beetle common in central and northern Europe, as well as Iceland and the island of Newfoundland. While native to Europe, it has been introduced to and is expanding its range throughout North America.

Black or dark piceous, upper surface more or less cupreous or greenish bronze, sides of prothorax, and often elytra, usually violaceous. Elytron with three rows of foveae and on each interval with suggestion of five ridges, so irregular and confluent as to give a scaly appearance. Length 21 to 26 mm. Use of *Carabus nemoralis* Muller as a biocontrol agent for multiple pests in large scale farming operations was tested (Lee and Edwards 1999). *Carabus nemoralis* is a polyphagous predator of crop pest (Symondson and Williams 1997).

2.3.6 Genus: ANCHOMENUS

Anchomenus dorsalis (Pontoppidan)

Body dorsally without metallic lustre, brownish red to reddish brown or reddish piceous, with lateral margins of pronotum and elytra paler, yellowish; tarsomeres paler, yellowish. Eyes moderately protruding. Antennae elongate, antennomere 9 more than four times as long as wide. Mandible not particularly elongate, with retinacular tooth covered by labrum in dorsal view. Anterior edge of mentum tooth not or very slightly emarginate. Submentum with two lateral setae on each side. Frons with two distinct rufous median spots in many specimens. Labrum with well impressed, more or less isodiametric meshes. Pronotum with well impressed moderately transverse meshes on disc, with isodiametric meshes at base between impressions; lateral margins with more

Fig. 27. *Anchomenus dorsalis* (Pont) (female) (https://en.wikipedia.org)

Fig. 28. *Anchomenus dorsalis* (Pont) (male) (https://en.wikipedia.org)

or less distinct transverse meshes. Elytra with well impressed, very transverse meshes (Figs 27 and 28).

In total, removal of *Anchomenus dorsalis* (Pont.) along with other carabids lead to an increase in aphid number. This indicates that ground beetle depressed aphid population in winter wheat field in Munich, Germany (Lang 2003). It is considered as an important aphid predator (Skuhravy 1959a, Sunderland 1975, Scheller 1984) and affected the establishment and early population growth of aphids (Sunderland & Vickerman 1980, Griffiths et al. 1985).

2.3.7 Genus: NEBRIA

***Nebria brevicollis* (Fab.)**

Pronotum wider than long narrowed towards base, with deep basal print. Elytra black, striae deeply punctured. Abdominal sternites with one bristle on each side of the median line (Figs 29 and 30). *Nebria brevicollis* (Fab.) was reported as a strictly predator of egg and larvae of *Helicoverpa* armigera (Hb.) in USA (Burgess et al. 2002) but it is a common predator of *Macrosiphum avenae* Theo in United Kingdom (Sopp and Wratten 1986).

Fig. 29. *Nebria brevicollis* (Fab.) (From https://en.wikipedia.org)

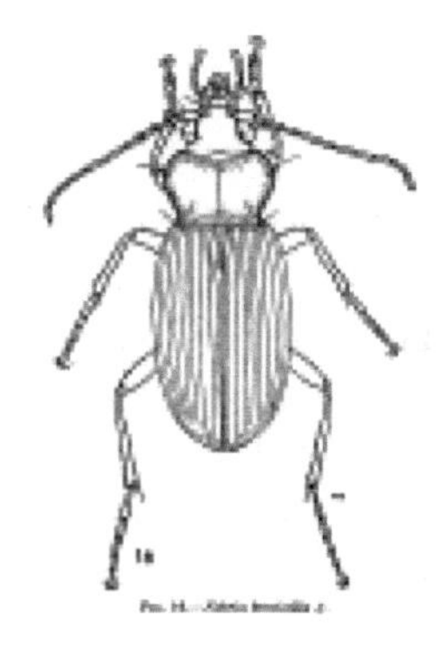

Fig. 30. Schematic diagram of *Nebria brevicollis* (Fab.) (After Carl H. Lindroth, *RES* 1974)

Nebria brevicollis (Fab.) is a polyphagous predator in Belgium (Loreau 1984, Symondson and Williams, 1997) but it is also common in agricultural fields (Lovei and Sarospataki 1990) and is a voracious predator with a wide prey range (Nelemans 1987a) in Europe. Author also collected *Nebria* sp. from grape vine in Israel during 2004 (Plate 1).

2.3.8 Genus: NOTIOPHILUS

***Notiophilus biguttatus* Fab.**

Small (3.5 to 5.5 mm), black to brassy coloured beetle with obvious protruding eyes and a parallel sided body form. The elytra possess two bright metallic, brassy yellow stripes whereas *Notiophilus substriatus* Water is totally metallic colour (Fig. 31). Adult beetle overwintered and breed in the spring. The next generations of adults were active in spring and early summer. The beetle is very widely distributed in arable crop and gardens and is also associated with woodland, as were many ground beetles. Some adults have wings, although, as is common in the ground beetle many do not.

Fig. 31. *Notiophilus substriatus* Water (From http://www.wikiwand.com)

This species is considered to be relatively mobile and dispersed through a large area. Larvae are probably subterranean, and pupation takes place in the soil. *Notiophilus biguttatus* (Fab.) is a rapidly darting, diurinally active polyphagous predator (Loreau 1984), consumed soil invertebrates, especially springtails and aphids on the ground.

2.3.9 Genus: CALOSOMA

Calosoma calidum (Fab.)

Size elongate to ovate, head triangular and shiny, labrum transverse and emerginate at apex, mandibles striate on upper surface and toothed at base only; fronto-clypeal suture not prominent and front bisulcate, antennae slender, pubescent from fifth joint, joint 2 about one-third as long as 3 and both are compressed; eyes large, lateral and prominent. Head not constricted behind the eyes (Figs 32 and 33).

Pronotum cordate, transverse, impunctate but granulated, posterior and anterior angle obtuse and median line present. Elytra elongate, ovate, basal margin not bordered, striae obsolete, disk granulated, leg long. *Calosoma calidum* Fab. can also eat living katydids, sting bugs, a cicada nymph, an ant, spiders, a butterfly, chrysalis, measuring worms, other larvae and sometime sowbugs (Slough 1940). The adult *Calosoma obsoletum* Say attacks and consumes living ranga caterpillars (*Hemileuca oliviae*) and grasshopper. Unrestained larvae of this insect in the field were observed feeding on range of caterpillar (Burgess and Collins 1917). The stomach content of *Calosoma*

Fig. 32. *Calosoma calidum* Fab. (male) (From https://en.wikipedia.org)

Fig. 33. *Calosoma calidum* Fab. (female) (From http://www.wikiwand.com)

calidum (Fab.) was examined and found them to have taken only animal food including grasshopper (Forbes 1883). The insect is a common predator of maize cutworm *Mythimna* sp whereas *Calosoma panagaeoides* Laferte predate on cowpea aphid in Karnataka, India (Kumar 1997).

Calosoma maderae var. _indicum_ Hope

The genus is large and elongate ovate about 22 mm long, head triangular and shiny, elytra, palpi, mandible, antennae and legs brownish black, labrum transverse and 5 to 6 marginated at apex, mandible striate on upper surface and toothed at base only; fronto-clypeal suture not prominent and front bisulcate, antennae slender, pubescent from 5th joint, joint 2 about one-third as long as 3 and both are compressed. Eye large, lateral and pronminent. Head not constricted behind the eyes. Pronotum cordate, transverse, impunctate but granulated, posterior and anterior angles obtuse and median line present. Antennae slender, elytral intervals conspicuously punctuate, surface uniformly granulated, humeral angle rounded. Scutellum small and transverse (Saha et al. 1992).

In India this species is distributed in Kashmir, Kumaon, Dehradun; Uttar Pradesh; Chapra and Pusa; Bihar and West Bengal. This species is the predator of caterpillar, grubs, maggots and other similar animals (Mani 1990). Both larvae and adult are predacious. *Calosoma maderae* Hope adults and larvae are nocturnal predator in maize field and hide in burrows in soil by day. Adults and larvae are seen to eat larvae and pupae of the cutworm *Mythimna separata* (Walk), in laboratory. They have modified forelegs that enable them to break into termite mounds, Isoptera, and also found to eat termites in laboratory.

2.3.10 Genus: ASAPHIDION

Asaphidion flavipes Linn

A small (4-5 mm), bronze coloured ground beetle with prominent eyes and with elytra decorated by prominent depressions and microsculpture (Figs 34 and 35). *Asaphidion* usually push their eggs into soil, ovipositor used to form ovigerous capsules for single eggs in the soil. The eggs lay at bottom of subterranean nests guarded by the mother until larvae are pigmented.

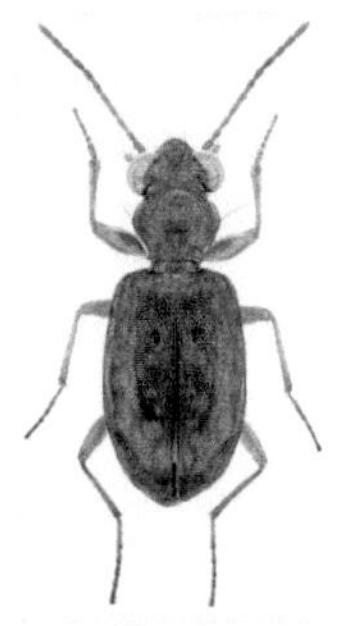

Fig. 34. *Asaphidion flavipes* Linn (From https://en.wikipedia.org)

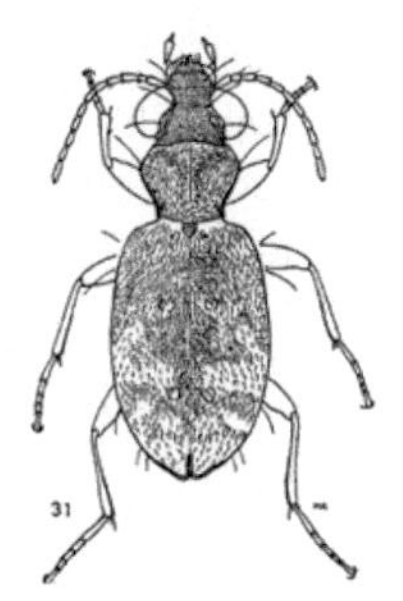

Fig. 35. *Asaphidion curtum* (Heyden) (After Carl H. Lindroth, *RES* 1974)

Ovipositor can be used to push egg into soil. This is a common predator of *Macrosiphum avenae* Theo in United Kingdom (Sopp and Wratten 1986) but it was recorded as polyphagous predator of different crop pests in Netherlands (Brandmayr and Brandmayr 1979) and Belgium (Loreau 1984).

2.3.11 Genus: TRECHUS

Trechus quadristriatus (Scrank)

Head rather narrow, polished integuments, with well evident microsculpture on the whole body, isodiametric on the head, transverse polygonal on pronotum, clearly transverse on elytrae. Head, pronotum and temples glabrous. Frontal furrows complete, regular, curved; front and hind edges of clypeus straight. Eye medium size, slightly longer than temples. Elytrae very wide, convex, glabrous and amply rounded at apex (pseudophysiogastric). Shoulders rounded. Elytral furrows thin and regular. Legs ferruginous, short and stout. First two tarsal segments of forelegs in males amply dilated, hooked on inner edge, with adhesive hairs on the lower face (Fig. 36).

Fig. 36. *Trechus quadristriatus* (Scrank) (From https://no.wikipedia.org)

Trechus quadristriatus (Scrank) is one of the most abundant small predatory carabids found on and around cabbage plots at Wellesbourne. The insect is dimorphic in wing length. Although the long-winged form of *T. quadristriatus* is capable of flight, few individuals appear ready for flight at any one time. The insect is freely mobile and active during the daytime and activity varies according to temperature but the lower temperature activity threshold for *T. quadristriatus* is less than 40° F. The distribution of the species on cabbage differ markedly. During the daytime the insect occur on the shaded ground under plants. Similarly, during the growth of a cabbage crop the number of this insect is directly correlated with the amount of plant cover present (Mitchell 1963). *Trechus quadristriatus* (Scrank) is a polyphagous predator of crop pest in Belgium but the appearence of this insect in the order grass-pine-beech, suggested carabid community organisation moving from succession to climax (Loreau 1984).

2.3.12 Genus: OXYLOBUS

Oxylobus porcatus (F.)

Head wide and smooth, a few short and deep striae intermingled with course punctures extending along the sides of neck constriction just outside the deep and parallel frontal impressions; continued in front to clypeal suture but posteriorly not reaching the neck constriction which is deep and prominent; labrum with median lobe truncate and well advanced, bearing a single pore; clypeus with well developed tooth on each side of labrum, clypeal suture distinct on middle and faint at extreme side; left mandible in male dilated at base and little upwards; eyes flat and conspicuous enclosed in genae. Prothorax (3.25 mm × 4.50 mm) a little wider than head and a little more than one and one third wider than long, sides parallel, a single setae on each side in front placed on inner edge of the border without breaking it; median line deep, basal favae obsolete, lateral channel deep and moderately rounding the front angle without forming which is moderately rounded transverse impressions rather shallow. Elytra (7.25 mm × 4.50 mm) as wide as prothorax and striate one and 3-5 longer than wide, 7-striate, with trace of another stria within marginal channel near apex, striae deep, strongly punctuate, punctures not increasing in size near apex; striae 1, 2 and 5 free at base, 3 joining 4; interval convex, 5 and 6 carinate on inner margins, 7 fine and strongly carinate, not reaching the base and joining 6 behind, 3 with a setirferous pore near the apex, epipleurae strongly punctuate. Prosternum moderately punctuate on sides, metepisterna strongly punctuate bordered on outer margin; basal abdominal segments with scattered punctures, last three segments with a row of punctures on sides and smooth on middle (Figs 37 and 38). The insect was first recorded from Nilgiri Hills of India (Hogan 2012). Among the different species *Oxylobus dekkanus* Andr was recorded as a predator of termite in India (Kumar 1997). *Oxylobus dekkanus* Andr and *O. punctatosulcantus* Chaudoir found in millet, maize, mulberry, eucalyptus plantations and mango orchards, trees, top fruit, forests, and woodland in India (Rajagopal and Kumar 1992).

Fig. 37. *Oxylobus* sp (From http://www.wikiwand.com)

Fig. 38. Elytral structure of *Oxylobus* sp (After Carl H. Lindroth, *RES* 1974)

2.3.13 Genus: CYMINDOIDEA

***Cymindoidea indica* (Schmidt-Goebel)**

Head oval with two supraorbital setae on either side, pronotum with two lateral setae on either side, antennae with three basal segments glabrous, elytra truncate at apex (Fig. 39), elytral epipleurae not interrupted by inner plica and first segments of anterior tarsi in male dilated (Saha et al. 1992).

Cymindoidea sp.was recorded as predator of termite in Kanataka, India (Kumar 1997).

Fig. 39. *Cymindoidea indica* (Schmidt-Goebel) (Images credit: Author)

2.3.14 Genus: RISOPHILUS

Risophilus atricapillus Linn

Head black and little narrowed behind. Body pale yellowish; mouth and thorax reddish; elytra obsoletely striated. Neck not apparent, thorax conidiform, a little longer than broad. Palpi filiform, terminated with thick ovoid truncate joint. Abdomen is very much depressed, tarsi with the fourth joint bifit (Fig. 40).

This insect was recorded as common predator of *Aphis fabae* Scop. in East Angila (Davies 1973).

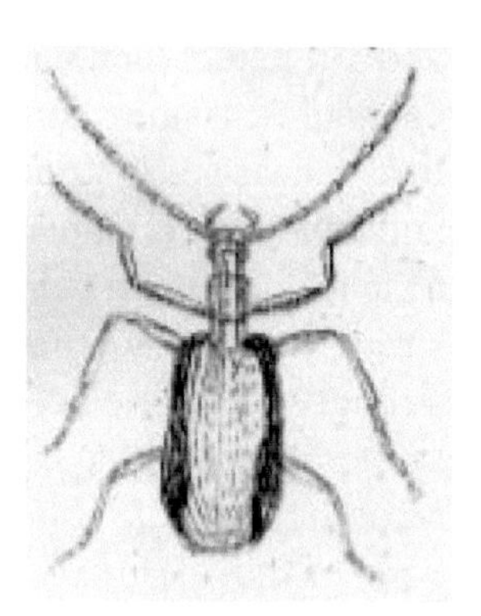

Fig. 40. *Risophilus* sp (Images credit: Author)

2.3.15 Genus: OMPHRA

Omphra sp

General body colour of adult beetle is black or brownish red; ligula, antennae and pulpi brownish red. Body setae black, brownish red or grey in colour. Legs are black, brownish red or dark red in colour. Head is convex, narrower than pronotum, widest between eyes, sparsely punctuate and setose on frontal foveae and inner side of eyes. Labrum transverse and short, apical margin with six setae; with median emargination, outer apical angles rounded, fine setae present on either side. Clypeus transverse, with 6-9 setae, apical margin emarginated frontoclypeal suture faint; frontal foveae depressed and setose; anterior and posterior supraorbital setae present, eye encircled by small setae. Genae slightly convex, asetose; neck glabrous, mandibles stout, exposed anteriorly, posterior covered by labrum, scrobe without setae. Mentum round and rectangular in shape, median tooth strongly developed and stout, pointed at apex, triangular, shorter than the lateral lobes, with 1-3 setae; lateral lobes strongly developed, obtuse at apex; base of mentum with 1-2 setae. Ligula small, elongate, tapered towards the apex, sides deeply depressed, with a median longitudinal channel, three pairs of setae present laterally. Palpi short, setose and 4 segmented; 4th palpomere of both maxilla and labium dilated or not dilated, maxillary and labial palpomere 2 as long as 4, 3 shortest. Antenna filiform, setose; scape elongate. Antennomere densely setose from one third of the base of antennomere 4, last antennomere oval or elongated oval.

Pronotum large, punctuate and setose, cordiform, almost round or not, widest at anterior third, strongly convex or slightly convex or flat; with a distinct shallow median line extending neither to apex nor base. Centre area asetose and glossy, densely punctuate and setose laterally. Anterior and posterior margins emarginated with a dense fringe of yellow short hairs, anterior or posterior transverse sulci faint; anterior angles round. Lateral margins distinctly or weakly sinuate in front of the posterior angles. Lateral margins with elongate setae from anterior margin to hind angle. Base narrower than apex, hind angles obtuse, base distinctly emarginated medially. Scutellum large, triangular, pointed, sparsely setose, variably punctuate. Elytra longer than pronotum, oval or intermediate between oval and oblong, convex, equal in with or widest before middle; humerus pronounced or not pronounced. Setae grey, black or brownish red. Interval strongly or weakly convex, with a row of black, brownish red or grey setae along on either side, intervals 3, 5, 7 and 9 with 2 to 6 randomly located elongate tactile setae; apex straight or obliquely truncate; emarginated or not emarginated. Sutural angle pointed, weakly emarginated, deeply emarginated or not emarginated. Venter sparsely punctuate and setose, except posterior area of prosternum and small median area of sternites 3-5. Prosternal process pointed. Hind wing absent (Raj et al. 2012). *Omphra* sp feed on termite and lepidopteran larvae infesting crop pests in India (Kumar & Rajagopal 1997).

Omphra pilosa Klug.

Body length 14-17 mm, colour black, coxa and trochanter brownish red, legs black, labrum transversely not extended, clypeus with 6-8 setae, mentum oval and rectangular, median tooth with two setae, base with two setae. Last maxillary and labial palpomere dilated. Last antennomere elongate oval. Pronotum strongly convex, entirely rounded, widest in middle. Lateral margins strongly sinuate to base (Fig. 41).

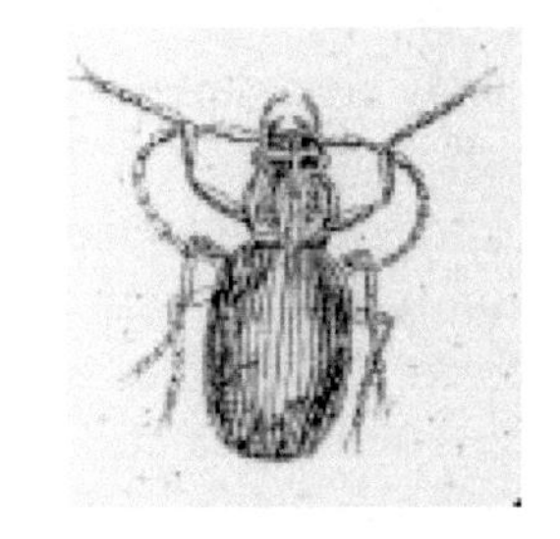

Fig. 41. *Omphra pilosa* Klug. (Image credit: Author)

Base slightly emerginate laterally; scutellum sparsely densely punctuate; elytra elongated oval, widest behind the middle. Humerus not pronounced. Intervals strongly convex, apical margin obliquely truncated with weak emargination (Raj et al. 2012). Among the other species *Omphra hirta* (Fab.), *Omphra pilosa* Klug, *Omphra atrata* (Klug), *Omphra complanata* Reiche, *Omphra rotundicollis* Chaudoir and *Omphra drumonti* Raj, Sabu & Danyang are predominant. The abundance of the preferred prey resources were ground surface dwelling ants and termites (Prasad and Rajagopal 1990), with subterranean habits of living in small burrows under large stones and protection by the odoriferous defense gland secretion might have allowed flightlessness to evolve in *Omphra* (Raj et al. 2012). Aptery was accompanied by cursorial adaptations, namely, elongated and strong legs in contrast to the slender and weak legs in other oriental helluonine genera, short median tooth of mentum and protruding mandibles for the partly subterranean way of life. The habit of storing dead ants in burrows for later use (Prasad and Rajagopal 1990), protection by the odoriferous defense gland secretions, cursorial adaptation, ability to live in burrows under rocks and idleness when faced with danger, might have contributed to the wider establishment of *Omphra* in the Indian subcontinent.

This insect is a predator of the teak defoliator, *Hyblaea puera* Cramer (Lepidoptera: Hyblaeidae) in an intensively managed teak plantation at Veeravanallur, Tamil Nadu (Loganathan and David 1999). *Omphra pilosa* Klug. (Coleoptera: Carabidae) appeared to feed exclusively on termites in the field. Adults are nocturnal, larvae are diurnal and store termites in burrows and were a potential predator on termites. Adults feed on all castes of termites in the field but preferred workers, a termite is consumed in 45 sec. Adults and larvae were found in Maize, Mulberry, Eucalyptus and Mango, orchards, fruit, trees, forest, woodland and cereals (Kumar and Rajagopal 1990). *Omphra pilosa* adults and larvae feed exclusively on termites including alatae, the larvae store termites in their burrows (Rajagopal and Kumar 1992).

2.3.16 Genus: AGONUM

***Agonum dorsalis* (Pont.)**

General body length 7-10 mm; colour metallic due to the bright green or bluish iridescence on the dorsal surfaces. The elongate habitus with slender appendages, distinctive colour (Figs 42 and 43), and the presence of 5-6 dorsal punctures with setae on each elytron make this species unmistakably different from other appalachian species (Fig. 43) of the genus (Lindroth 1966). In the Nearctic Region, the genus is represented by 72 species. About 10 of them inhabit GSMNP (Ball and Bousquet 2001). The insects lay their eggs at bottom of subterranean nests guarded by the mother until larvae are pigmented, ovipositor used to form ovigerous capsules for single eggs in the soil, constructed mud cells for single eggs, beetle collect mud on tip of abdomen for this. *A. dorsale* even dips abdomen into water first to aid mud construction, thought to protect against desiccation and entomogenous fungi, pathogens, disease, and natural enemies. *Agonum dorsalis* (Pont.) is a common predator of *Tetraneura nigriabdominalis* (Sasaki) in United Kingdom (Sunderland 1975, Edward et al. 1979) but it was a polyphagous predator of crop pest in Netherland. (Brandmayr and Brandmayr 1979). Among the other species *Agonum cupripenne* Say, *A. gratiosum* Mannerheim, *A. melanarium* Dejean, *A. placidum* Say and *A. puncticeps* Casey were predominant in USA (Esau 1968) and *Agonum albipes* (Fab.); *Agonum obscurum* (Paykull); *Agonum assimile* (Mots.); *Agonum quadripunctatum* (Buck.), *Agonum versutum* Sturm, *Agonum*

Fig. 42. *Agonum dorsalis* (Pont.) (From https://en.wikipedia.org)

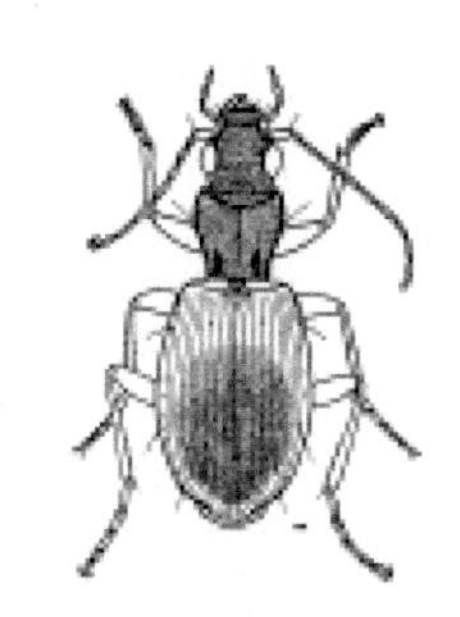

Fig. 43. *Agonum dorsalis* (Pont.) (After Carl H. Lindroth, *RES* 1974)

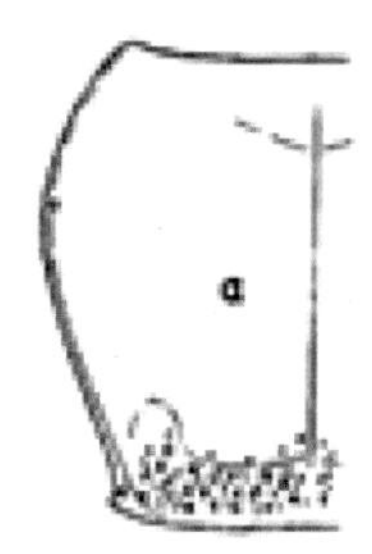

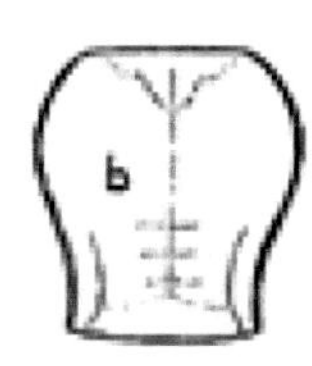

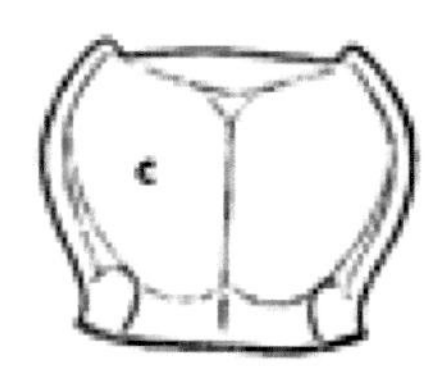

Fig. 44. *Agonum*. Pronotum of species (a) *albipes*; (b) *obscurum*; (c) *assimile*; (d) *quadripunctatum* s. str. (After Carl H. Lindroth, *RES* 1974)

viduum Panzer and *A. thoreyi* Dejean (Figs 44 and 45) were found from the record of Lindroth (1974).

Agonum placidum Say. was found by Rivard(1964a) to be mostly on open ground with dry soil. Lindroth (1966) indicated that the species was found among dead leaves and similar material, and shady places near water of rivers and lakes. Enough dead vegetation provided adequate cover for the species. Lang (2003) reported that *Agonum mulleri* was common predator of wheat aphid like *Sitobion avenae* Theo., *Metopolophium dirhodum* (Walk.) and *Rhopalosiphum padi* Linn. in Munich, Germany.

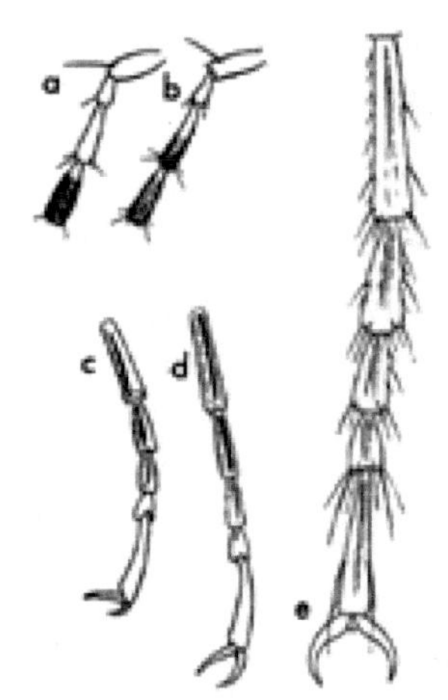

Fig. 45. *Agonum*, Antennal base of (a) subg. *Agonum* (b) subg. *Europhytus*. Hind-tarsus of (c) *versutum* (d) *viduum* (e) *thoreyi*. (After Carl H. Lindroth, *RES* 1974)

2.3.17 Genus: CALATHUS

***Calathus fuscipes fuscipes* Goeze**

Robust with a body length of 10-15 mm; recognized by presence of setigerous punctures on both third and fifth interval; the first antennal segment pale, legs piceous to clear red. Both *Calathus fuscipes* Goeze and *Calathus melanocephalus* (Linn) and *Calathus erratus* (C.R. Sahlberg) (Fig. 46) were recorded from apple orchard in Central Hungary. Beside this *C. ambiguous* (Paykull) and *C. micropterus* (Dufts.) (Fig. 47) are found from the record of Lindroth (1974).

Fig. 46. *Calathuserratus* (C.R. Sahlberg) (From http://www.wikiwand.com)

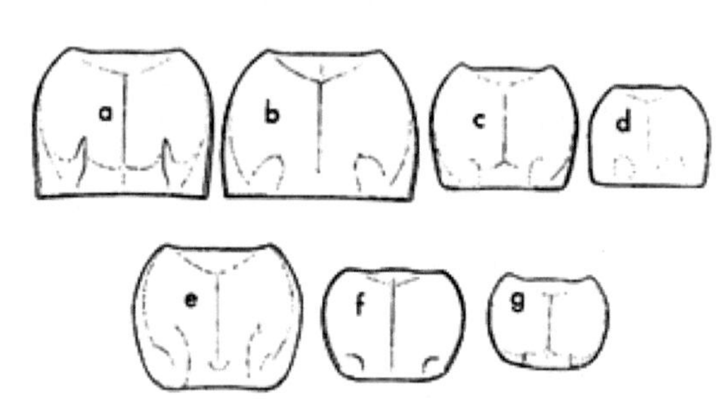

Fig. 47. Pronotum of (a) *Calathus erratus*; (*b*) *ambiguous;* (*c*) *micropterus*; (f) *Synuchus nivalis*; (g) *Olisthopus rotudatus* (Courtesy: Carl H. Lindroth, *RES* 1974)

Adults emerged in June and July. For overwintering, they grouped together under a stone or under the soil surface. They reproduce in the next spring. Their lifespan is more than one year. Females laid down eggs each year. After each egg production old female's ovarioles are filled with a structure called yellow body. Using mud, females built a small cell to protect the eggs. Adults feeding regime was mixed zoophagous and phytophagous. It preyed upon slugs, and other invertebrates. An individual ingested one time its own weight daily (0.062 g) (http:// www.sibnefl.eu/gb/ Coleoptera/Carabidae/img100/ eco100.HTM). Briggs (1957) reported that *Calathus fuscipes* Goeze is a polyphagous predator of crop pests. Wilkinson (1965) attributed the failure of introduced cinnabar moths to establish a population at Addotsford, British Columbia to predation of newly formed pupae by this insect.

Calathus gregarius Say

Body glossy, rufo-piceous underneath, above nigro-piceous. Palpi and antennae pale testaceous; prothorax quadrangular with the angle rounded; anteriorly emarginated; channeled; basilar impressions slight; lateral margin rufous dialated posteriorly, elytra furrowed, furrows impunctured; four punctiform impressions between the second and third furrows, the three anterior one adjacent to the latter, and the posterior one to the former; legs pale testaceous, claw

Fig. 48. *Calathus mollis* (Marsham) (From http://www.wikiwand.com)

pectinated (Fig. 48). *Calathus gregarius* Say was known from Nova Scotia to Iowa and south to Florida and Texas (Esau 1968) which was also recorded in the leaf litter of a balsam-spurce forest on Isle Royale (Adams 1909) as well as pastures and a meadow in New York (Wolcott 1937) from grass, logs, and a meat trap in the Mt. Desert area of Maine (Procter 1946). Six specimens of *Calathus gregarius* Say were dissected by Forbes in 1883 and found to have consumed caterpillars and other insect larvae, the remaining one-third of their food being pollen of grasses.

2.3.18 Genus: PTEROSTICHUS

Pterostichus sp

Adult beetle large (12 to 18 mm), stout, black; pronotum narrower than the elytra having a long sharp mandible, elytra with two small puncture (Fig. 49). Casey (1918) reported that the species like *Pterostichus stygicus* Say as being abundant and found from New York and Pennsylvania to Indiana in USA. The distribution of *P. chalcites* was recorded by Leng (1920) and Casey (1918) to be both Atlantic and Pacific. Procter (1946) recorded *Pterostichus luctuosus* Dejean from under stones and around ponds on the Mt Deseret island of Maine. The distribution was completely trans-American, being found from Newfoundland to Vancouver Island, British Columbia, in the north, and south as far as Virginia and west into Washington and Oregon. Rivard (1964a) reported occurrence of this predator on moist soil, with moderate to dense vegetation. The insect can well be an important factor in the natural biological control of rootworms in Iowa. The species is numerous enough in most fields so that it can certainly be number one predator on these larvae. Starved laboratory specimens have been observed to immediately pounce upon corn rootworm larvae when presented to them.

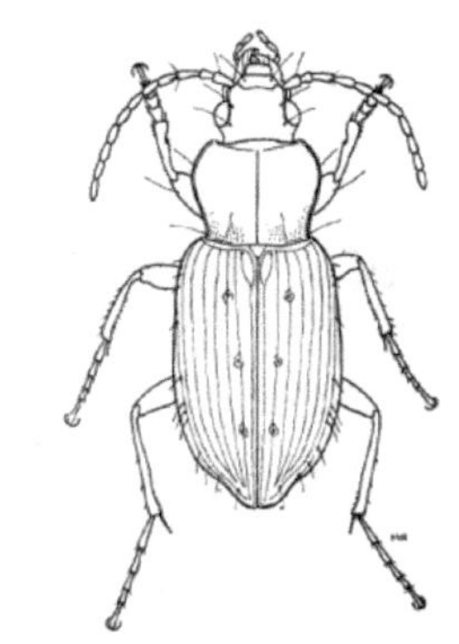

Fig. 49. *Pterostichus* sp (After Carl H. Lindroth, *RES* 1974)

Wilkinson (1965) attributed the failure of introduced cinnabar moths to establish a population at Addotsford, British Columbia for predation of newly formed pupae by *Feronia (Pterostichus) melanaria* (Illiger). Lindroth (1966), however, indicated that the distribution was decidedly eastern, reached only as far west as Indiana, and in Canada only into Ontario and not further west. This insect was recorded as common predator of *Aphis fabae* Scop. in East Anglia and Belgium (Briggs 1965, Davis 1973, Loreau 1984, Hance 1987). The species is very hygrophilic but is also rather eurytopic, being found near sandy waters and marshes hardwood forests, or adjoining meadows with high vegetation where the vegetation is rich and frequently contained mosses (Lindroth 1966). The author also stated that the distribution of *Pterostichus patruelis* Dejean and *P. femoralis* Kirby were trans-American, but mostly northern, reaching south as far as Kansas. In contrast *Pterostichus lucublandus* Say was the common predator of armyworm (Lepidoptera: Noctuidae) in no-till corn in USA (Laub and Luna 1992) whereas *Pterostichus niger*, *Pterostichus cupreus* Linnare are polyphagous predator (Symondson and Williams 1997). *Pterostichus chalcites* Say, a metallic green coloured beetle was recorded almost exclusively in arable land in the prairies, USA. Rivard (1965a) found the species in or near woods with moist soil. *Pterostichus melanarius* (Illiger) a predator of winter wheat pest in Germany (Lang 2003) which is a nocturnal species, shelters under rocks and vegetation in the day and completely ground based (Heyler et al. 2009). In addition to this *Pterostichus anthracinus* (Illiger) was assessed in the European Red Lists prepared by the IUCN for the European Commission was also included as predator of crop pest (Fig. 50).

Fig. 50. *Pterostichus anthracinus* (Illiger) (From https://en.wikipedia.org)

2.3.19 Genus: ABACIDUS

***Abacidus permundus* Say**

Colour typically dark and without conspicuous patterns, but often with a strong sheen like polished metal (Fig. 51). They are widely distributed and inhabit a wide range of terrestrial habitats. Unlike the more basal ground beetles which only consume small animals, the Pterostichinae include a large proportion of omnivorous or even herbivorous taxa. Blatchley (1910) indicated that *Abacidus permundus* Say was found frequently throughout the southern half of Indiana, beneath logs in open and sandy woods (http://www.wikivisually.com/wiki/Abacidus).

Fig. 51. *Abacidus permundus* Say (Image credit: Author)

This insect was taken in wild rye wasteland, a blue grass roadside and a sweet clover field in Kansas, USA (Walkden and Wilbur 1944). The observation on limited laboratory feeding trials indicate that this species was a capable predator (Esau 1968).

2.3.20 Genus: CYCLOTRACHELUS

***Cyclotrachelus (Evarthrus) sodalis* LeConte**

Colour black, legs usually black sometimes red; penultimate article of labial palpus plurisetose (usually) or bisetose; pronotum rectangular to cordate, basal lateral fovea of pronotum bistriate, monostriate, or a single puncture, always distinctly impressed; basal lateral seta of pronotum on lateral bead or beside it; elytron with seventh interval usually raised at base, 1-5 punctures on medial side of third interval; hind wings absent; metepisternum short, with lateral margin equal in length to anterior margin; article five of tarsus usually with a row of setae on each ventrolateral margin; venter impunctate, usually slightly rugose; females with two setae on last sternum of abdomen; eversion of internal sac of median lobe of male genitalia usually to right, less often dorsoapical, and rarely to left (Fig. 52). Larva Pterostichini; antenna with five articles; urogomphi short, terete, curved toward eac. All of the species are flightless: not only are the hind wings of all individuals atrophied, but the metathorax is reduced, and the elytra are fused along the suture. It is not surprising, therefore, that geographical variation was marked, that most of the species had restricted ranges, and that closely related species were often allopatric—facts which indicated restricted powers of dispersal. Members of the genus inhabited deciduous forests or open country. Those species which occur in open places was of northern and western in distribution. Conversely, the ranges of the more numerous forest species was generally southern and east of the Mississippi River. Little is known about the biology of this genus (http://nature.berkeley.edu/).

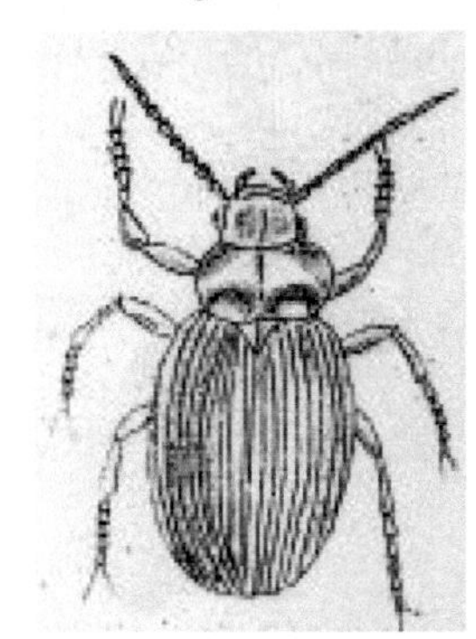

Fig. 52. *Cyclotrachelus sodalis* LeConte (Image credit: Author)

Probably the members are omnivorous, as are most Carabidae. Spores of fungi in the gut of *C. faber* Germar, and ant remains in the gut of *C. sodalis* LeConte. The species was reported from Pennsylvania, Ohio, Indiana, Iowa, Nebraska, Missouri, Kansas, Texas and Louisiana (Esau 1968). Dissection of this specimen revealed that none of the food was of vegetable origin. Another specimen contained traces of algae in addition to several insects (Forbes 1880). Seven more specimens (Forbes, 1883) had consumed about 7% vegetable food and the remainder caterpillars, Scarabaeidae, Coleoptera larvae, cankerworms and other insects. Slough (1940) found his captive specimen quickly took animal food, sometimes even fighting over mosels while a third specimen appropriated the tidbit. Live food eaten by the beetles included grasshopper nymphs, leafhoppers, stinkbugs. Freitag (1968) reported the species to eat ants.

2.3.21 Genus: LORICERA

Loricera pilicornis (F.)

Small (6-8.5 mm) bronze-black coloured ground beetle with conspicuously bristly antennae (Fig. 53). Very common everywhere but especially gardens, open gravelly or sandy sites and streamsides in mountains. In European works, this species was found to avoid flooded habitats, but to be present in moisted ones. Northeastern France results indicated a discrepancy with that pattern. Up to now, it has been missing from all investigated moisted environments, flooded or not. It was very eurytopic, largely diurnal species, but with a preference for open habitats. Widespread in most habitats, including mountains, where it occurred along stream banks, and on moraine or summit breccia (http://www.sibnef1.eu/gb/ Coleoptera/Carabidae/ img167/ eco167.HTM). A single individual per stem controls aphid density 50 per stem in patch (Bryan and Wratten 1984). This insect is also a polyphagous predator of crop pest in Belgium (Loreau 1984) and USA (Symondson Williams 1995).

Fig. 53. *Loricera pilicornis* (Fab.) From https:// en.wikipedia.org

2.3.22 Genus: POECILUS

***Poecilus cupreus* (Linn.)**

Adults 9 to 13 mm long, typically bright metallic green in colour but may also be bronze or coppery coloured. The first two antennal segments yellowish or orangish in colour. The insect is a pterostichine carabid with a wedge – shaped body well adapted to pushing and burrowing its way through soil and leaf litter (Fig. 54).

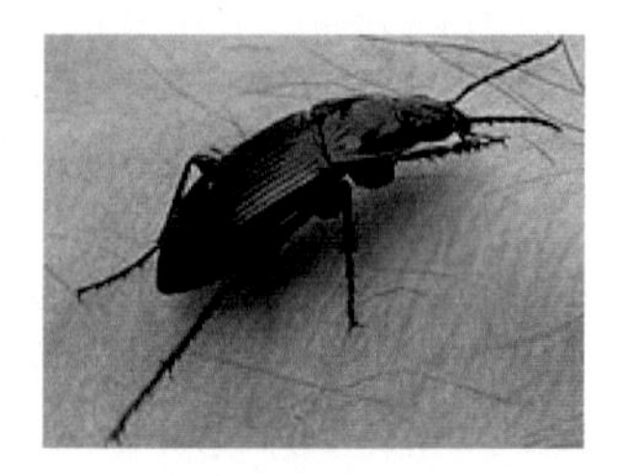

Fig. 54. *Poecilus cupreus* (L.) (From https:// en.wikipedia.org)

Poecilus cupreus (L.) is common in agricultural fields throughout Europe where it prefer relatively moist habitats. The larvae and the adults are polyphagous and consume a variety of prey. *P. cupreus* are typically spring breeding carabid species that overwinter as an adult. Both adults and larva consume insect prey, although in England the adults also damage young plants of the genus *Beta* such as beetroot, chard and sugarbeet. *P. cupreus* (L.) is a generalist carabid beetle with no associations known to particular prey types. Common in all field crops, this beetle can consume fly eggs and pupae, aphid and soil mite. *P. cupreus* thrive in tall, dense vegetation and is often found close to water course. Hedge rows and uncropped areas around crops provide an overwintering habitat for many predatory carabid beetle species, including *P. cupreus* (Heyler et al. 2009). In total, removing carabid beetles including *Poecilus cupreus* (L.) led to an increase in population of *Sitobion avenae* Theo, *Metopolophium dirhodum* (Walk.) and *Rhopalosiphum padi*

Linn. indicated that ground beetle depressed aphid population in the winter wheat field in Munich, Germany (Lang 2003).

2.3.23 Genus: ACUPALPUS

***Acupalpus (Egadroma) smargdula* (Fab.)**

Adult small, elongate-ovate, impunctae except the basal area of pronotum, dorsal side glabrous, antennae pubescent from segment 3. Head and elytra blackish brown, pronotum reddish brown (Figs 55 and 56). Labrum straight, transverse; clypeus straight, transverse with a pair of setae antero-laterally; 1st segment of maxilary palpi pointed at apex; head glabrous; antennae brown, slender 1st and 2nd joints glabrous, rest hairy; eyes moderately large, black, coarsely facetted. Pronotum oval, margined laterally, median line fine, posterior angle of pronotum coarsely punctured. Among the different species *Acupalpus dubius* Schil., *A. exigus* Dejean, *A. consputus* Dufts., *A. meridianus* (Linn.), *A. elegans* Dejean; *A. dorsalis* (Fab.) were found from the record of Lindroth (1974) (Figs 57 and 58).

Fig. 55. *Acupalpus meridianus* Fab. (From https://en.wikipedia.org)

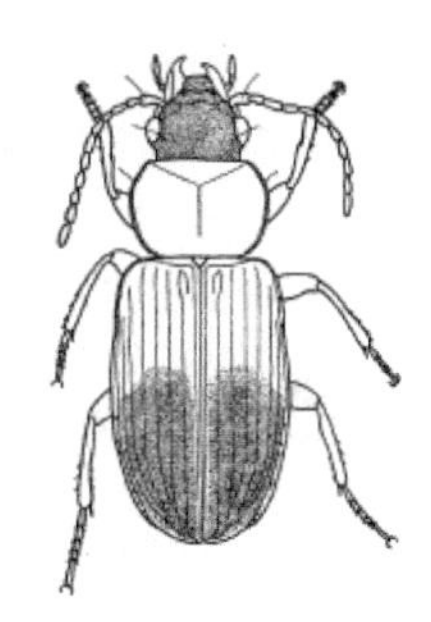

Fig. 56. *Acupalpus* (Egadroma) *smargdulus* Fab (After Carl H. Lindroth, *RES* 1974)

Elytra laterally margined, humeral angle well developed, striate punctate, apical margin with few setae, legs brown. Body length varied from 5.5-6.0 mm (Lindreth 1974).

The insect is usually attracted to light. It is generally found in the months of October, November and March in Eastern India. Chiu

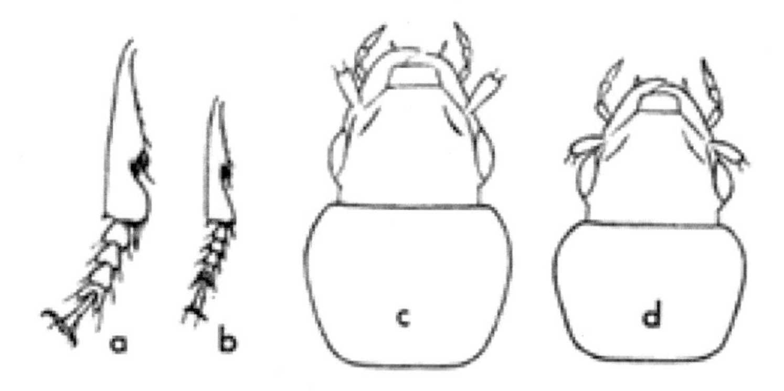

Fig. 57. *Acupalpus*. Front tibia of male in (a) elegans, (b) dorsalis, Forebody of (c) *dubius*, (d) *exigus*. (From Carl H. Lindroth, *RES* 1974)

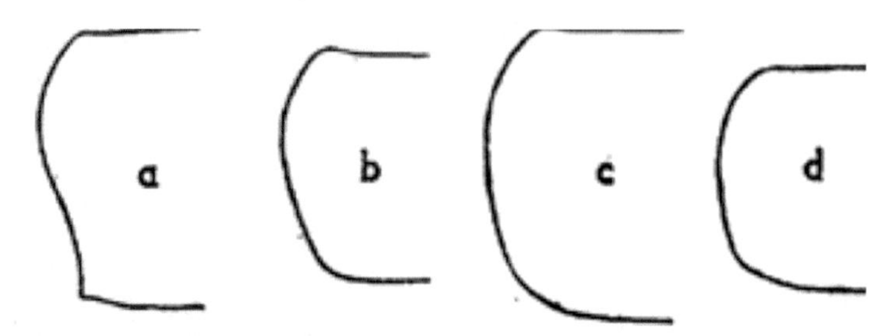

Fig. 58..*Acupalpus*. Pronotum of (a) *consputus*, (b) *meridianus*, (c) *elegans*, (d) *dorsalis* (From Carl H. Lindroth, *RES* 1974)

(1979) also reported that *Acupalpus inornatus* Bates is a predator of brown planthopper (*Nilaparvata lugens* (Stål.) and other leafhopper of rice in Taiwan.

2.3.24 Genus: GNATHOPHANUS

Gnathaphanus licinoides **Hope**

Elytra with intervals 3, 5 and usually 7 with dorsal punctures conspicuously enlarged and impressed, colour black, leg black (Fig. 59).

This genus is represented by 11 species and is distributed in South East Asia, Australia and New Guinea (one species). Six species are found in India, of which *Gnathaphanus licinoides* Hope is a common predator of larvae of potato cutworm (Kumar & Rajagopal 1997).

Fig. 59. *Gnathaphanus melbournensis* (Castelnau) (From https://nl.wikipedia.org)

2.3.25 Genus: PHEROPSOPHUS

Pheropsophus occipitalis **(Mac Leay)**

Pheropsophus aequinoctialis (Linn.), one of the first ground beetles, was described from the Neotropic ecozone. It was first described by Carl Linnaeus in his 1763 work Centuria Insectorium, under the name *Cicindela aequinoctialis.* It was also later described by Johan Christian Fabricius in 1775 as *Carabus complanatus* and by Guillaume-Antoine Oliver in 1795 as *Carabus planus*, both of which were junior synonyms of *Pheropsophus aequinoctialis* (Linn.). It was later made the type species of the genus *Pheropsophus* (George 1996). The subtribe Pheropsophina is one of four subtribes of Brachininae (Coleoptera, Carabidae, Brachininae) (Erwin 1970, 1971, Lorenz 2005a, b). The beetle of *Pherosophus occipitalis* (Mac Leay) moderately large and elongate, head brown with rectangular black patch on vertex and one elongate black patch on front; eyes black; palpi and mandible deep brown; clypeus and labrum not so deep brown, antennae filiform, scape dilated, pedicel smaller than all joints. Pronotum black but with two brown elongate patches to the lateral sides of the posterior part; elytra black with one pair of brown patch on the base of elytra near the humeral angles and the other elongate brown patch on the sides of lateral margin centrally, legs brown but hind femur with blackish colour at the apex. Elytra elongate oblong, apices truncate, three abdominal segments exposed, humeral angle rounded striate broad and broader than intervals, hairy intervals raised, shiny, scutellum triangular and wrinkled, length 16 mm to 18 mm (Fig. 60).

Fig. 60. *Pheropsophus occipitalis* (Mac Leay) (Image credit: Author) (See Plate 2)

Female *P. aequinoctialis* (L.) significantly prefer to lay eggs in sand with mole cricket tunnels compared with artificially created tunnels or sand without tunnels. Physical tunnel presence influences oviposition depth, but is not the only factor influencing oviposition. *Pheropsophus aequinoctialis* (Linn.) lays clutches of 25–60 eggs close to the burrows of mole crickets. The eggs are white and rectangular in outline, with rounded apices. The three instars (larval stages) which follow are white, with a cream head capsule, and darker colouration at the tips of the mouthparts. The larvae of *Pheropsophus aequinoctialis* (Linn.) are specialized predators of *Scapteriscus* mole cricket eggs in some parts of south America. First instar larvae of this predator are very active and have long legs enabling them to find the mole cricket egg chamber underground. Once the first instar penetrates an egg chamber it begins to feed and soon after will molt into a very different looking second instar. The second instar and all later instars have legs that are much shorter and wider body when compared to the first instar. Finally, after all of the eggs are consumed, the larva moults in to the third and final instar and is confined to the egg chamber until the adult emerged and digs its way out.

Scapteriscus mole crickets deposit 25-60 eggs depending on female size and age in small, ovoid egg chambers. The egg chambers are located 9-30 cm underground depending on soil moisture and are closed off from their main tunnels (Forrest 1985). The pupa was of the typical form in ground beetles (Frank et al. 2009). The larval stage of *Pheropsophus aequinoctialis* (Linn.) is a specialist predator of *Scaperiscus* mole cricket egg recorded that larvae of *Pherosophus aequinoctialis* (Linn.) developed under laboratory conditions on a diet restricted to mole cricket eggs (Orthoptera, Gryllotalpidae). Habu and Sadanaga (1965, 1969) reported that larva of this beetle could recognize that large numbers of mole cricket eggs were adequate for their growth and development. The authors also reported that this insect predator could be used as an additional biological control agent in the vicinity of water bodies, on their banks, in particular, and could be beneficial in integrated pest management because application of chemical pesticides is prohibited from use in such habitat. The insect feed on *Neocurtilla* and *Scapteriscus* mole cricket eggs in the laboratory. Under normal temperature (25°C) egg, larva and pupal stages varied from 13.5, 25.9 and 20.4 days respectively. Sex ratios of emergent adults were not substantially different from 1:1 (Frank et al. 2009).

Adult *Pherosophus* has a crepitating behaviour like other Brachinini, producing quinines (Zinner et al. 1991) and they were nocturnal, running on sandy trails or reverie beaches, hiding during day under stones, grass, clumps, and drift logs and often in aggregations; they were predatory on other insects and also consume some plant materials, such as ripe fruits of *Astrocaryum* sp., apalm (Reichardt 1971). This insect is the common predator of cricket (*Gryllotalpa* sp.) eggs. Among the other species *Pheropsophus hilaris* var *sobrinus* Dej. is reported as a predator of the larva of rhinoceros beetle *Oryctes rhinoceros* Linn. in manure heaps and pits in south East Asia (Atwal 1986). It was also found that beetle predated on fruit fly (*Dacus* sp.) maggots and pupae falling on the ground in guava field in India (Mahamed Jalaluddin 1999). Effect of field boundary habitats on early colonisation of rice fields by *Pherosophus javanus* was recorded in Philippines. (Sigsgaard et al. 1999). Adults of *Pheropsophus aequinoctialis* (L.) (Coleoptera, Carabidae, Brachininae, Brachinini), are largely nocturnal predators and scavengers on animal and plant materials. The daily food consumption of a pair of adults is the equivalent to 1.2-2.3 large larvae of *Trichoplusia ni* (Hübner) (Lepidoptera, Noctuidae). Larvae developed under laboratory conditions on a diet restricted to mole cricket eggs (Orthoptera, Gryllotalpidae); none survived under any other diet offered, thus they were specialists. Adult *Pherosophus aequinoctials* feed on adult *Scateriscus* mole crickets in sand-filled containers in the laboratory. Most neonate larvae need to be in a cell or pit of sand (or earth) resembling a mole cricket egg chamber before they will feed on mole cricket eggs. The behaviour of *Stenaptinus jessoensis* (Morawitz) as adult and larva is very similar to that of *P. aequinoctialis* except that adults are mainly diurnal.

In the southeastern United States, South American *Scapteriscus* mole crickets are serious pests of turf and pasture grasses and vegetable seedlings. *Pheropsophus occipitalis* Macleay and *P. lissoderus* are considered as the important predators of *Oryctes rhinoceros* (Linnaeus). As these predators help in the natural check of the pest population, conservation of the predator fauna is essential. Among the different species, *Pheropsophus hilaris* (Fab.), *Pheropsophus lissoderus* Chaudoir, *Pheropsophus occipitalis* (Macleay), and *Pheropsophus (Stenaptinus) andrewesi* Jedlicka are predominant in India (Park et al. 2006, Kirschenhofer 2010). The adult *Pheropsophus occipitalis* (Mac Leay) was also recorded by author from Thailand during 2016.

2.3.26 Genus: BRACHINUS

Brachinus favicollis Erwin

Small to medium sized beetles with blue or brown elytra and ferrugineus head and thorax, second antennal segment short and all segments with at least some pubescence; eyes prominent and with a "bead" each at dorsal edge; labrum short and broad; mandibles each with a single tooth; galae palpiform and two-segmented, with apical segment obconical; maxillary palpi four-segmented, with second and third segments obconical; labial palpi three-segmented, with first and second segments obconical; anterior tibiae with an "antennal comb" (formed of a fringe of stiff setae) in middle of posterior edge, and with an apical tibial spine; elytra costate to a variable degree, and truncated to dehiscent at apex; epipleura elevate above margin of elytra from humeri to outer apical corner; ventral thorax and abdomen mostly pubescent; middle coxal cavities disjunct; abdomen with seven visible sterna; parameres of male genitalia glabrous, reduced, the right one being very much smaller than the left (Fig. 61).

Fig. 61. *Brachinu* sp (From https://en.wikivisually.com)

Brachinus crepitans (Linn.) can reach a length of 7–10.2 millimetres (0.28–0.40 in), with an average of 8 millimetres (0.31 in). Head and pronotum are brown, while elytrae are greenish (Fig. 62). When disturbed, the species shoot liquid from two glands through their anus. Since one of the glands contains hydrogen peroxide and the other hydroquinone, when the two contents mix with enzymes in a "firing chamber", the liquid explodes, and harms the attacker. *Brachinus favicollis* Erwin spend the dark hours patrolling the ground for other insects that it could eat. The beetle often laid their eggs directly on a potential food source such as a rotting animal corpse (Erwin 1965).

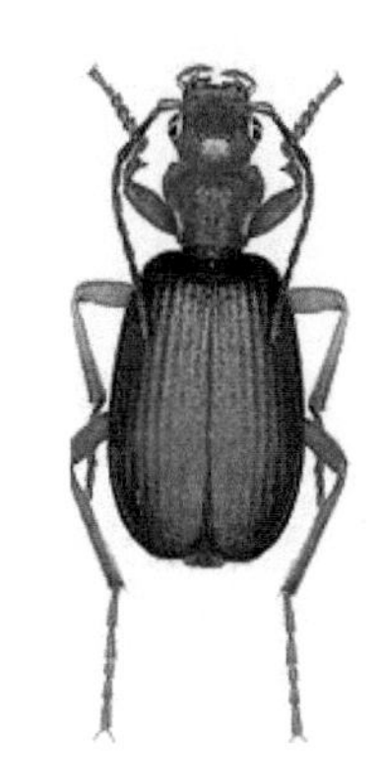

Fig. 62. *Brachinus crepitans* (Linn.) (From https://en.wikipedia.org)

When the eggs hatch into a larva it can then utilize the source of food for weeks as it develops. The larvae also eat the larva of other beetles and insect that are adjacent to it (available at Quirk Brain. com). Similar to other beetles in the Carabidae family, bombardier beetles hunt for their food with their smaller insects being their prey but when this predator is not hunting for prey, it will congregate with others in damp, dark places (Poetker 2003). The author also described that any place will do for a ground beetle to lay its eggs, so long as it's out of the way of most predators but not too far away from a good food source. Small underground tunnels or cracks in rotting wood are viable places, as are the decomposing remains of other living things (which quite often serve as the food source). The egg hatched into the larval stage, which began taking in nourishment from the food source and occasionally moulting. After it shed its skin for the last time, it metamorphose into a pupa, the stage at which the juvenile looks most like the adult which it will eventually become. At the end of the pupal stage, the pupa sheds its skin and a new adult bombardier beetle emerged.

Beetle quite naturally share some of the habits of its family, and like most other ground beetles, tend to come out at night to prey on smaller insects. This carnivorus beetles find its prey in trees and on the ground. With their ideal habitat being on riverbank, another moist area, it is where mating and ovipositing takes place. Bombardier beetles are ectoparasitoids for some species, it has been found that their larvae developed on the pupae of whirling beetles (Gyrinidae) and water scavenging beetles (Hydrophilidae) (Saska and Honek 2004, Riddick 2008). The beetle had the ability to aim and shoot a potent chemical from its internal chambers deterring the attack of spider, mantids and ants (Juliano 1985).

2.3.27 Genus: LYMNASTIS

***Lymnastis* sp**

Adult *Lymnastis pilosus* Bates small and elongate (2 mm) beetle, head, pronotum and elytra light brown; antennae, leg also light brown, labrum and clypeus straight, a pair of big punctures on front, antennae short and stout, 1st joint glabrous, rest hairy, eyes small. The elytral colour of *Lymnastis galilaeus* Piochard (Fig. 63) is mostly identical to *Lymnastis pilosus* Bates. Pronotum cordate, transverse, lateral margin slightly sinuate at base, posterior angles rectangular, anterior angles obtuse, disc slightly hairy (Figs 63 and 64). Elytra elongate and parallel sided, fine striate finely punctuate (Saha et al. 1992).

Fig. 63. *Lymnastis galilaeus* Piochard (From http://carabidae.org/)

The body of *L. schachti* completely depigmented, surface including all appendages light yellow, pronotum and elytra somewhat translucent. Head comparatively narrow and elongate, neck thick, head widest immediately at base. Labrum six-setose, with very deep, semicircular excision that in middle leaves only a small margin. Clypeus and frons in middle convex, posterior to lateral part of clypeal suture with a large, fairly groove on either side. Laterally of this groove and just about eye with a shallow carina that posteriorly extend to near posterior margin of eye. Eye comparatively small, laterally very little projected, about as long as distance between posterior border of eye and base of head. Eyes composed of 50 ommatidia. Vertex dorsally slightly convex, without transverse impression. Mandible short, strongly incurve toward apex. Labium in middle without a distinct tooth, though with a very shallow convex only. Preapical palpomere rather globular, densely pilose. Antenna rather short, surpassing base pronotum by about three antennomeres. Median antennomeres slightly less than 1.5x as long as wide. Only one elongate supraorbital seta present.

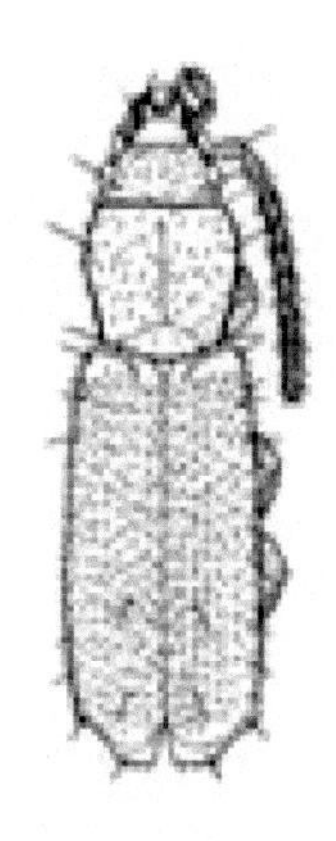

Fig. 64. *Lymnastis* sp (From Carl H. Lindroth, *RES* 1974)

Frons on either side behind clypeal suture with a postclypeal seta. Dorsal surface with isodiametric to slightly transverse microreticulation that becomes very superficial in middle, and with very sparse puncturation and comparatively elongate, erect pilosity. Pronotum, comparatively narrow, not much wider than long, dorsally rather convex, widest at about anterior two fifth. Apex very gently concave, apical angles rounded, not produced, lateral margin gently rounded in anterior two thirds, very faintly concave near basal angles which bear a minute denticle. Basal angle about 100°, base in middle strongly lobate, laterally excised, lateral parts of base slightly oblique. Apex and base not bordered, lateral margin with narrow margin throughout. Median line distinct, not reaching apex, basal transverse sulcus deep, very oblique, laterally ending in the rather deep basal impressions. Both marginal setae elongate, anterior seta situated in front of widest diameter, basal seta set slightly inside of basal angle. Surface with about isodiametric microreticulation and with rather coarse and with fairly dense puncturation. Anterior and lateral margins furnished with a fringe of rather short setae that are directly horizontally, surface with moderately elongate, erect to slightly inclined pilosity, and close to apex with a row of more elongate, erect setae.

Elytra narrow elongate, almost twice as long as wide, widest at about apical third, though elytra but little widened towards apex. Humeri produced though convex, lateral margin almost straight but faintly oblique. Lateral part of apex characteristically excised, median part straight and slightly oblique, with a distinct angle between median and lateral parts. Apex deeply cleft in middle. Transparent part of apex unusually wide. A similar short of elytral apex has been never detected in any other species. Surface very weakly though almost fully striate, intervals depressed, striae impunctate, though intervals each with a row of coarse punctures and short, erect setae. Lateral and apical margins with a fringe of rather short setae that are directed horizontally; 3rd interval with two elongate fixed sitae, both situated in apical two fifth of elytra. Lateral margin apparently with a row of four anterior and 4 posterior marginal pores and setae, that is widely interrupted in middle, though all setae broken. Surface with about isodiametric, rather superficial microreticulation. *Lymnastis* sp was recorded as a predator of sugarcane pest in Malayasia (Lim and Pan 1980).

2.3.28 Genus: BEMBIDION

Bembidion quadrimaculatum (Linn.)

Head with two supraorbital setae on each side, frontal furrows moderately deep and not curving round behind the eyes; the mandibles with a seta in the scrobe, ligula bisetose, setae placed close together; paraglossae hardly extended beyond the ligula, apical joint of palpi exceedingly small and sublate, penultimate joint dilated and pubescent, antennae with the first two joints glabrous (Figs 65 and 66); elytra nine-striate, scutellary striote as a rule rather slightly impressed, apical recurrent striole absent; outer margin of from tibiae straight from base to apex (Saha et al. 1997).

Fig. 65. *Bembidion lampros* (Herbst.) (From https://species.wikimedia.org)

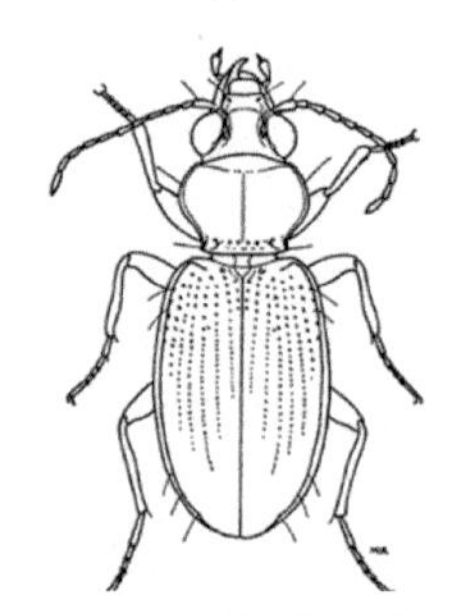

Fig. 66. *Bembidion lampros* (Herbst.) (After Carl H. Lindroth, *RES* 1974)

Although this predator is distributed around the world but *Bembidion eutherum* Andrewes, *Bembidion xanthracum* Choudoir, *Bembidion* (*Notaphominus*) *foveolatum* (Dejean) and *Bembidion kara* Andrewes are very common in India (Park et al. 2006). Among the other species *B. lampros* Herbst and *B. quadrimaculatum* are the most abundant predators during the early establishment phase of aphid population and significant inverse correlations are found between peak number of *Rophalosiphum padi* Linn. per shoot and numbers of predators from these taxa in Sweden during 1981 and 1982 (Cheverton 1986) and also found to prey on *Hylemya* eggs in Manitoba (Lindroth 1963). Feeding trials in the laboratory showed that the adult *B. quadrimaculatum* can eat corn rootworm larvae, and could definitely be explored as a biological control agent. The predators were frequently seen running on the surface in the cornfield in the hot summer, even during daylight hours, rushing from one cravices to another, thus seemingly on a constant search for food (Esau 1968). This insect is also recorded as the most numerous ground beetle predacious upon cabbage maggot eggs in the field (Pitre and Chapman 1964). During laboratory trials they found this ground beetle consumed 67% of the eggs offered. *Bembidion propernus* and *B. lampros* were recorded from apple orchard in central Hungary (Horvatovich and Szarukan 1986). *Bembidion obtusum* Serville was also recorded as a predator of *Macrosiphum avenae* Theo in United Kingdom (Stopp and Wratten 1986).

Bembidion sobrinum Boheman

Small, elongate about 4.5 mm, head, pronotum and elytra excepting the apical portions all dark black, antennae excepting the 1st joint, palpi, legs dark brown and eyes black. Head triangular, finely granulated, a pair of furrows on the frons, three pairs of brown setae on labrum; labrum transverse; clypeus emerginate, eyes prominent, bulging, finely facet, pronotum cordate, transverse, front margin wider than basal margin, sinuate at basal margin, median line fine, anterior angle obtuse (Saha et al. 1997). Elytra elongate-ovate, shoulder prominently developed, eight rows of striation, fine, minute punctures on the striae, extends upto apex, elytral apical part brown (Fig. 67).

Fig. 67. *Bembidion laevigatum Say* (From https://nl.wikipedia.org)

Among the different species *Bembidion semilunium* Neto and *Bembidion sobrinum* Boheman have been recorded as important predators of Brown Plant Hopper and other leafhoppers in Taiwan (Chiu 1979, Li 1983) whereas *Bembidion lampros* Herbst is a predator of *Myzus persicae* Sulz. in United kingdom (Dunning et al. 1975). The daily consumption of *Myzus persicae* Sulz. at 24°C is 6.7 in USA (Tamaki and Olsen 1977). Author also recorded the *Bambidion* sp. from a grassy land at Taiwan during 2016 (Plate 2).

2.3.29 Genus: DIORYCHE

Dioryche cuprina (Dejean)

Species medium size and elongate about 7.5 mm in length. Head, pronotum and elytra dark brown, pulpi, labrum, antennae except 1st and 2nd joints, leg deep brown; eyes black. Labrum straight, somewhat rounded, three pairs of brown setae present anteriorly, apically sparsely and finely punctured. Clypeus slightly emarginated and short. Head glabrous, antennae long, slender, hairy and last segment pointed at apex, eyes large and lateral.

Pronotum somewhat cordate, transverse, impunctate except the basal margin and posterior angle which finely and closely punctured, basal margin and front margin nearly equal, lateral margins slightly sinuate at base, posterior angle obtuse, median line not distinct. Elytra elongate and sinuate at the apical end, glabrous, striate, a good numbers of coarse dots on the intervals 2 and 4 humeral angle well defined. Scutellum triangular and transverse (Fig. 68). Among the other species *Dioryche colembensis* Nietner is a common predator of *Aphis craccivora* Koch. in India (Bhat 1984).

Fig. 68. *Dioryche cuprina* (Dejean) (From https://en.wikiwand.com)

2.3.30 Genus: LEBIA

Lebia (Poecilothais) calycophora Schmidt-Goeble

Lebia (Poecilothais) calycophora Schmidt-Goeble is a common beetle in India which resemble the bombardier-beetles quite closely in size and colour, but which may be distinguished by the comb-like form of the tarsal claws 4.3-5.0 mm. Reddish brown or orange, mat; labrum, palpi, antennae, lateral explanate parts of pronotum, and legs yellowish brown; elytra generally yellowish, rarely somewhat reddish, lateral area (intervals 7 to 9) sometimes slightly brownish, blackish longitudinal fascia situated on suture, widened near base, occupying intervals 1 to 3 or 1 to 4, somewhat narrowed at anterior one-third (intervals 1 and 2), thence a little widened (occupying intervals 1 to 3) behind middle, black oval patch situated behind middle on intervals 5 to 8, black fascia and black patch generally not coalescent with each other, but interval 4 interposed between them; ventral side yellowish brown, sometimes more or less reddish (Fig. 69). The species is sometime confused with *Lebia darlingtoniana* Baehr. Distributed in Myanmar (Burma), China (Fijian), Indonesia (Sumatra), India, Japan, Laos, Malaysia, Thailand, Taiwan (Formosa), Vietnam (http:// carabidae .org/taxa/calycophora-schmidtgoebel).

Fig. 69. *Lebia darlingtoniana* Baehr, (From https://en.wikipedia.org)

It has also been reported more often than any other insect as destroying the Colorado potato beetle (Comstock 1960) and other crop pests (Allen 1958) in USA. Among the other species *Lebia vittata* Fabricius was recorded from Dekalb cornfield, Cayler hill and Cayler bench of USA (Esau 1968) but Madge (1967) stated the distribution in United States and Canada.

2.3.31 Genus: CALLEIDA

***Calleida splendidula* (F.)**

The adults of *Calleida splendidula* (F.), 7 to 10 mm long, 2.5 to 3.5 mm wide, slender, head and elytra green or blue-black, mentum supported at base by projecting submentum. Head with two supraorbital setae on either side, antennae dark glabrous with three basal segment lighter, pronotum with two lateral setae on either side, thorax and legs yellowish-red with tips of femora and tarsi dark, elytra truncate at apex (Fig. 70). Males with a double row of papillate hairs on the undersurface of the first three protarsal segments and first two metatarsal segments (appearing white), but female tarsi with pubescent (straw-coloured) (McWhorter et al. 1984).

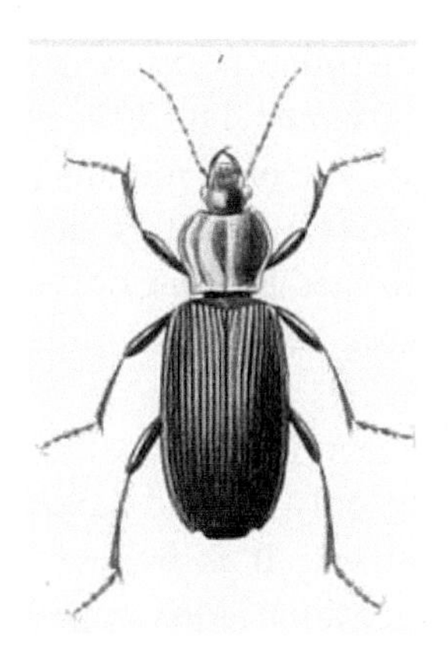

Fig. 70. *Calleida amethystina* (Fabricius) (From https://en.wikipedia.org)

Both head and thorax of *Calleida amethystina* (Fabricius) are golden yellow which can differentiate the insect from other species (Fig. 70). Caged adult females lived an average of 230 days with a mean preovipositional period of 11 days and laid an average of 800 eggs. The eggs are round, white, semi-opaque, approximately 0.75 mm in diameter, covered with sand particles, and attached by a silken thread to a leaf or other available surface such as a stem or twig. While an egg is still held by the abdominal tip, the female cover it with sand or dust particles, and bind it with silken thread to form a purse. The "egg purse" is attached to a leaf by a silken thread. Developmental times at 22 to 28°C for eggs, larvae, and pupae are approximately 4 to 6, 12 to 18, and four to six days, respectively. Larvae are predaceous except while undergoing sclerotization following hatching and moulting. They are highly cannibalistic and must be reared in individual containers. They feed readily on lepidopteran eggs as well as small larvae.

On soybeans, populations are estimated as high as 5400/ha in Gadsden County, Florida and 9600/ha in Alachua County, Florida. *Calleida decora* (F.) adults and larvae have been observed feeding on velvetbean caterpillar, *Anticarsia gemmatalis* Hübner; cabbage looper, *Trichoplusia ni* Hübner; soybean looper, *Pseudoplusia includens* (Walker); and other lepidopterous larvae (Whitcomb and Bell 1964, McCarty et al. 1980, McWhorter et al. 1984). Over 10% of the total insect predation (almost 20% during one season) of *Anticarsia gemmatalis* Hubner larvae (1st to 4th instars) artificially placed on soybean foliage was by *Calleida decora* (Fab.). Of the 21 predation observations involving *Calleida decora* (F.) during a total of 4 seasons, 19 were by larvae and 2 by adults. Adult *Calleida decora* (Fab.) were found confined in small field cages on potted soybeans, consumed an average of 6.4

small (1st to 3rd instars) *Pseudoplusia includes* (Walk.) larvae/24 hr (Richman et al. 1980). Among the different species *Calleida pallipes* Andrews and *Calleida splendidula* (Fab.) are common predators of the coconut leaf eating caterpillar, *Opisina arenosella* Wlk. in India. During larval development it consumed 11-13 caterpillars, adults consume one caterpillar per three days and live 6-14 months (Park et al. 2006, Pillai and Nair 1990).

The insect is also recorded from termite colony in Karnataka, India (Kumar 1997). Consequently *Callieda* sp. feed on larvae and pupa of coconut black headed caterpillar in India (Kumar & Rajagopal 1997). Among the other beetles *Calleida decora* (Fab.) is a small arboreal ground beetle, predaceous both as larva and adult. Common on various cultivated crops, it was apparently the only carabid have to complete its larval development on Florida soybean foliage. It is believed to be a major factor in suppression of velvetbean caterpillar *Anticarsia gemmatalis* Hubner on soybeans (Neal 1974). Predation by *Calleida decora* (F.) (Coleoptera: Carabidae) on velvetbean caterpillar (Lepidoptera: Noctuidae) was also studied in soybean, where live eggs and larvae were readily accepted as prey in USA (Fuller 1988).

2.3.32 Genus: SCARITES

Scarites subterraneus Fab.

Length of *Scarites subterraneus* Fab. 30 mm; breadth across the elytra 12 mm; buccal fissure extending beyond the base of mentum, head with one supraorbital seta, maxillae hooked at apex, elytra one and half times as long as wide, parallel sided (Fig. 71).

Fig. 71. *Scarites* sp (From https://upload.wikimedia.org)

Among the other species *Scarites subterraneus* Fab. is the common predator of caterpillar and wire worm in Eastern North Carolina, USA as per the record of north Carolina State University. This insect is a natural enemy of armyworm (Lepidoptera: Noctuidae) in no-till corn in USA (Laub and Luna 1992). From observations of limited feeding trials in the laboratory, an adult *Scarites quadriceps* Chaudoir, when unfed for a period of time, pounced upon a rootworm larva presented to it and immediately kill it. During a 24-hour period an adult *Scarites quadriceps* Chaudoir consumes several rootworm larvae. The predator also take adult corn rootworms and various lepidopterous larvae, some almost as long as the ground beetle (Easau 1968).

Scarites (Parallelomorphus) indus Oliv.

These beetles share physical characteristics of the more tropical stag beetles, but are not closely related. *Scarites* could often be found under loose rocks and boards. If touched, they often "play dead" by folding in their legs and arching their backs. The beetle large and elongate, head, pronotum, elytra mandible, labrum, clypeus black; whereas palpi, antennae and leg deep reddish black. Mandible bidentate; labrum trilobed, each lobe with setigerous punctures, clypeus transverse, front broad, with a pair of depressions and wrinkled. Head somewhat squarish, glabrous and sparsely punctured on vertex. Antennae short, stout and moniliform, hairy joints 1-3 glabrous, rest hairy. Eyes small and lateral. Pronotum somewhat quadrate, transverse, impunctate and glabrous, lateral margins bordered and constricted at base, basal margin somewhat truncate. Elytra elongate and parallel-sided, shoulders squarish, striate-punctate, some pores present on third stria and intervals convex. Elytra deeply striate, interval convex, size not less than 16 mm (Saha et al. 1997). Likewise *Scarites indus* Olivier was recorded as the predator of termite from Karnataka, India (Kumar 1997). The adult beetles are predators and has been observed overpowering mealworms much larger than themselves. (https://en.wikipedia.org/wiki/Scarites)

***Scarites punctum* Wiedem**

Species moderately large and elongate. Head finely and closely striate, densely punctuate behind elytra superficially striate, intervals flat, size not more than 14 mm, head, pronotum and elytra black; palpi, mandible, labrum, antennae joints 1-3, legs dark reddish black. Mandible unidentate, labrum trilobed, each lobe with setigerous punctures; clypeus transverse and slightly undulated; fronto-clypeal suture prominent, front broad with a pair of depressions; apical segment of palpi almost pointed. Antennal joints 1-3 glabrous and rest hairy; a pair of supraorbital setae present; eyes small and lateral; vertex coarsely and not deeply punctured. Pronotum subquadrate, transverse, lateral margins bordered, a pair of supraorbital setae present; eyes small and lateral; vertex coarsely and not deeply punctured. Pronotum subquadrate, transverse, lateral margins posteriorly, median line prominent from apex to base, posterior angle carinate, anterior angle obtuse (Saha et al. 1997). Elytra elongate and parallel sided, well developed humeral angle, striate-punctate, three large pores present on third stria (Fig. 72).

Fig. 72. *Scarites* sp with its prey (Image credit Franclsco Rodriguez (Faluk) (From http://www.biodiversidadvirtual.org)

Among the other species *Scarites aterrimus* (Morawitz), *Scarites molossus* (Banninger), *Scarites guerini* (Chaudoir), *Scarites silvestris* Laporte, *Scarites terricola* Bonelli, *Scarites subterraneus* F., *Scarites striatus* F. and *Scarites buparius* (Forster) distributed in Japan, Zambia, Colombia, Brazil, France, USA, Saudi Arabia and

Algeria respectively (Hogan 2012). *Scarites punctum* Wiedem is recorded as predator of citrus pest in Eastern India. Author also recorded some *Scerites* sp. from an important beach of Saudi Arabia during 2011.

2.3.33 Genus: MACROCHEILUS

Macrocheilus sp.

Size large about 15.00 mm, elongate-oblong, head rectangular, deeply and coarsely punctuated; apical palpi truncate, mandibles concealed by labrum, labrum glabrous, smooth, shiny, broad and semicircular; clypeus transverse and truncate, anteriorly with a row of punctures; fronto-clypeal suture prominent, antennae short and stout, hairy; eyes large, lateral and prominent. Pronotum cordate, transverse, deeply and coarsely punctured with erecy hairs, prominent median line, anterior angle rounded and posterior angle rectangular. Elytra elongate and parallel I shaped slightly truncate, striate-punctate, presence of seven striae and small scutellary striole, intervals convex with a irregular deep punctures, disk hairy, humeral angle rounded, scutellum moderate, apical angle rounded, punctured. Abdomen finely punctured, pro, meso and metasternum sparsely and deeply punctuate; legs hairy (Fig. 73). Among the different species *Macrocheilus impictus* Wied was predominant in termite colony in India (Kumar & Rajagopal 1997) whereas *Macrocheilus niger* was the predator of *Pareuchaetes pseudoinsulata* in Ceylon (Perera 1981).

Fig. 73. *Macrocheilus tripustulatus* (Dejean) (https://en.wikipedia.org

2.3.34 Genus: OPHIONEA

Ophionea indica (Thunberg)

Body length 6.73 to 7.13 mm, width 1.84 to 1.94 mm, antenna 10 segmented, characterized by the colour pattern of the elytra including the reddish-yellow base, median black fascia, and reddish-yellow apex. Two small whitish spots on each elytron also characterize this group. The aedeagus was rather simple in structure where the apical and basal orifices were in an unusual position probably because of a twist of the aedeagal median lobe. Genitalia: Aedeagal median lobe elongate, somewhat depressed laterally, with two surface sclerites on right side of apical orifice turned to left in left-ventral view with semisclerotized membrane and thin membranous part on left side; apical lobe short, peculiarly twisted at base and abruptly bent dorsad (Fig. 74).

Both the grubs and adults of the carabid beetle were important predators of brown planthopper (See Plate 2) (Samal and Misra 1984). *Casnoidea*

Fig. 74. *Ophionea indica* (Thunberg) (Image credit: Author)

Fig. 75. *Ophionea indica* (Thun.) inside a damaged sugarcane stem. (Image credit: Author)

indica (Thun.) was also found in rice fields preying on brown planthopper *Nilaparvata lugens* (Stål), Hemiptera, Delphacidae in India (Rajagopal and Kumar 1992). It was found to maintain good predator prey relationship between carabid beetle and brown planthopper (Bonn and Kleinwachter 1999). In Eastern India the adults of this insect was found to take shelter either inside mature rice stems or the sugarcane damaged by the lepidopteran borers (Fig. 75). The predators inside the rice stem were forced to come out during threshing of rice crop in farmyard. The population density of *Ophionea indica* (Thun.) fluctuated enormously among three different management approaches at different days after transplanting. The numbers of predators ranged from 4.0 to 7.25 per five double sweep nets in natural control approach. Comparatively reduced numbers of predators were found in need based control approaches which ranged from 1.75 to 5.75, minimum number of carabid beetles were present in the field treated with schedule based chemical control approach where numbers of carabid beetles varied from 1 to 4.5.

The relationship of brown planthopper and *Ophionea indica* Habu were determined by using correlation and regression analysis (Fig. 76). Brown planthopper and the predator population was significantly correlated in natural

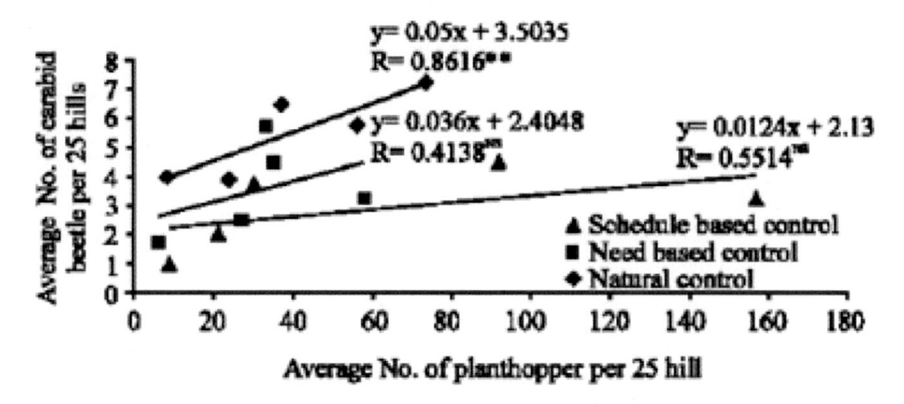

Fig. 76. Relationship between number of brown planthopper and Carabid beetle in different management approaches.

control approach ($r = 0.8616^{**}$, $y = 0.05x + 3.5035$). In case of need based control (NBC), brown planthopper and carabid beetle were weakly correlated ($r = 0.4138$ NS, $y = 0.036x + 2.4048$) while in schedule based approach, there was a poor correlation between brown planthopper and carabid beetle population ($r = 0.5514$ NS, $y = 0.0124x + 2.13$). The survival of brown planthopper and carabid beetle was affected by application of insecticides both in the field of need based and schedule based chemical control (Ullah and Jahan 2004). Among the other species *Ophionea interstitialis* (Scmidt-Goebel) was also reported to be a common predator of hopper pest in India (Park et al. 2006). Besides this midge galls were dissected (8523 galls from 52 sites) to look for *O. indica* larvae. Up to 3 larvae per 100 galls was found with highest numbers in the wet zone during the monsoon, extra hole in the gall suggested that the beetle larva detected the midge pupa within the gall and bored a hole to enter the gall and attacked the pupa. Wounded or dead gall midge pupae were found in galls where the predator is present (Kobayashi 1995).

Ophionea ishi hoashii Habu

General body length 6.39 to 6.68 mm, width 1.79 to 2.01 mm, the lack of fine pubescence on the head and prothorax. The lack of dorsal setae on elytral interval V was also a diagnostic character of *O. hoashii*. The beetles are active hard-bodied insects, elytra reddish-yellow or orange in apical area; with posterior whitish spots, but without anterior spots; elytral interval III with more than five dorsal setae proepisterna glabrous; metatarsomere IV bilobed. Prothorax shiny, less than twice as long as wide, with lateral setae on each side; lateral borders of prothorax much reduced posteriorly. The aedeagus is rather simple in structure where the apical and basal orifices are in an unusual position probably because of a twist of the aedeagal median lobe (Fig. 77). Both the shiny black larvae and reddish brown adults actively search the rice canopy for leaf folder larvae.

Fig. 77. *Ophionea ishi hoashii* Habu (Image credit: Author) (See Plate 2)

This beetle can be found within the folded leaf chambers made by leaf folder larvae. The predator larva pupate in the soil of wet land rice bunds on dry land fields. Each voracious predator consume 3 to 5 larvae per day, leaving only the head capsules. The adult also preyed on planthopper in rice field in Asia (Heinrichs 1994). Author also estimated the ground beetle population in rice field at International Rice Research Institute, Philipppines and found that

** denotes highly significant

Ophionea ishi hoashii Habu was more likely to survive in fields where non-inversion (e.g., chisel plough) tillage was used during land preparation.

2.3.35 Genus: EUCOLLIURIS

Eucolliuris olivieri (Buquet)

This species is unlikely to be confused with any other ground beetle. It is generally around 6.5 mm long, with a long head and pronotum, but typical in form otherwise. Coloration is key for distinguishing it from other members of the genus. The head and pronotum black, and the antennae light brown except the basal portion which is lighter. The elytra black with a brownish spherical mark at the tip. The legs light brown (Figs 78 and 79).

Antennal segment I short, less than half of head width across eyes; labium truncate with less than six setae, elytra not hairy. Pronotum and head densely punctate dorsally, laterally and ventrally, posterior lateral margin behind the eye relatively straight and oblique. Heinrichs and Barrion (2004) reported *Colliuris* sp as a pest control agent of rice crop in West Africa. Among the different species *Colliuris pensylvanica* Linn. was found in many habitats, but preferred wet, marshy areas. They could be found among leaf litter, rotting logs, and the ground among vegetation where they searched for prey (http:// data1. insectmuseum .org). This insect was also a possible aphid predator in USA, and Canada (http://bugguide .net/node/view /10161). Everly (1938) found the species under debris in a sweet corn field in Ohio, overgrazed pasture and an old alfalfa stand in Kansas habitats (Walkden and Wilbur 1944), from Ohio bluegrass sod fencerows (Dambach 1948), on vegetation over wet ground around marshes (Ball 1960). Open, dry ground with sparse vegetation was common habitat of this insect (Rivard 1964b). The distribution of this insect included Canada, Massachusetts, Indiana, Michigan as far south as Florida, and west to California (Leng 1920), Iowa (Jaques and Redlinger 1946), Indiana (Blatchley 1910) and was found in litter beneath logs, along fencerows, and in woods.

Fig. 78. *Eucolliuris olivieri* (Buquet) (Image credit: Author)

Fig. 79. *Eucolliuris olivieri* (Buquet) (Image credit: Author) (See Plate 3)

***Eucolliuris* sp.**

Beetle is generally around 6.0 mm long, with a long head and pronotum. Head subquadrate, finely closely punctate; mandibles without etigerous punctures in the scrobes, sides elevated; basal joint, antenna with a long bristle at apex; antennal segment I short, less than half of head width across eyes; front labrum slightly curved inwardly (almost straight) margin with setigerous punctures; clypeus with a puncture on each side, labium truncate with less than six setae, elytra not hairy, antenna and mouth parts testaceous. Pronotum and head sparsely punctured, shiny black head with fine punctations dorsally and ventrally, unpunctated laterally, lateral side of the head behind the eyes smooth and rounded; antenna dark reddish brown with yellowish segment I, II, basal two third of III and basal one half of IV, elytra dark brown, apices of femora black.

Pronotum subquadrate, one-third wider than long, wider than the head, widest at about the middle, apical angles narrowly rounded and slightly produced, lateral margin with a etigerous puncture on the apical third and another at the basal fifth, basal angles broadly rounded, marginal bead extending the side of seration of the head to the scutellar area, a median longitudinal pressed line, somewhat well marked, extending from the apex to near the base, with indistinct oblique lines, extending backwards from this median line. Scutellum triangular, apex narrowly rounded. Elytra with angles rounded margins not interrupted, surface with indications of faint strise and appearing as if covered with fine scales giving them a silky appearance, possibly granulate punctate, submargin with etigerous punctures, apical angles rounded. The legs are a light amber colour (Fig. 80). Front tibia with two elongate spines at upper end of the margination (the outer one apparently bifid for about two-thirds its length), calspin somewhat long; joints of the front tarsi shorter than those on the other legs, first joint slightly longer than the next two taken together and about as long as the apical joint, joints two, three and four gradually shorter, hind tarsi with the first joint, slightly longer than the apical joint, other joints shorter, claws distinctly shortly pectinate; hind tibia with spines similar to those on the middle tibia, on the outer side (http://bugguide .net / node/view/10161). It was also recorded as pest control agent of rice crop in West Africa (Heinrichs and Barrion 2004).

Fig. 80. *Eucolliuris* sp. (Image credit: Author) (See Plate 3)

2.3.36 Genus: ODACANTHA

***Odacantha (Odacantha) melanura* Linne**

Body slender, elongate varies from 6 to 10 mm, shiny and glabrous. Head elongate, labrum straight and transverse; front sparsely and finely punctured; vertex devoid of punctures, antennae slender, long, hairy 3rd joint onwards; eyes lateral and prominent; well developed neck in between head and pronotum, pronotum subcylindrical, finely and closely punctured.

Prothorax subcylidrical but more lengthy, devoid of lateral borders and whole body glabrous, elytra ovate-elongate, interval flat, head narrowed behind to a condyliform neck, prothorax subcylindrical, with more or less obsolete lateral borders. Elytra ovate-elongate, striate-punctuate, punctures fine and close; intervals not flat, humeral angle well developed, elytral apices obliquely truncate. Legs slender, long and not hairy (Fig. 81). The genus *Odacantha* Paykull is represented by *O. graciliceps* Bates and *O. punctata* Nietner (Saha et al. 1992) which are also recorded as predator of hopper pests in South Africa (Heinricks & Barrion 2004).

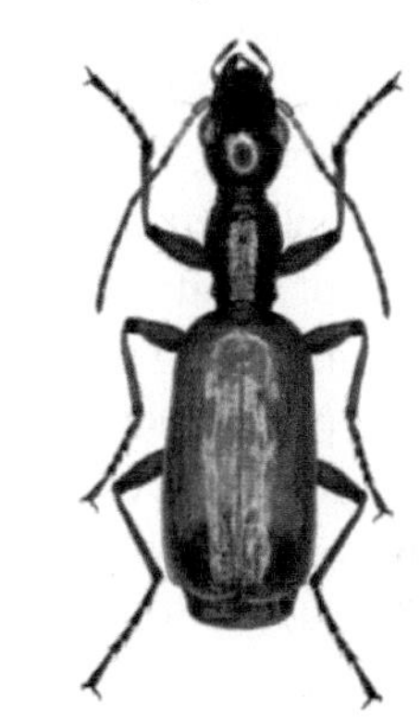

Fig. 81. *Odacantha melanura* Linne (From:https://commons.wikimedia.org)

2.3.37 Genus: DRYPTA

Drypta japonica (Bates)

Head and prothorax brown, elytra black along sutural, basal and lateral margins with a brown band covering ridge 3 to 7; head less punctuate behind the eyes, pronotum slightly longer than broad and densely punctuate; elytral punctuations on the intervals and none on the ridges; brown labrum trilobed with lateral lobes rather acute lateral; apices of femora I to III reddish brown; apical one half to one third of scape blackish brown; margins of frons in front of eyes converging apically; apico-lateral corner of elytra rounded; length 7-9 mm (Figs 82 and 83). The insect was recorded as predator of rice insects from Phillipines (Henrichs et al. 1994). Huge population of *Drypta japonica* (Bates) were found to take shelter inside the matured rice stem which ultimately came out during threshing of rice at the farm yard during the month of December in Eastern India.

Fig. 82. *Drypta japonica* (Bates) (Image credit: Author) (See Plate 3)

Fig. 83. *Drypta japonica* (Bates) on a rice stem. (Image credit: Author)

2.3.38 Genus: DIPLOCHEILA

Diplocheila sp.

Large in size, elytra, pronotum and head black, broad, rather quadrate, ocular ratio 1.47-1.50, piceous, shiny, microsculpture very faint isodiametric with scattered punctulae visible at >60 X, glabrous except for normal pair of supra-orbital setae over each eye, one seta at anterolateral comer of clypeus and four setae on labrum, clypeus, margin of labrum, maxillary and labial palps and antennae paler, rufopiceous, sometimes slightly infuscated but never as dark as vertex of head; labrum deeply and asymmetrically emarginate, right side more prominently produced; mandibles broad, asymmetrical, right smooth, left with prominent dorsal protuberance; labial and maxillary palpomeres fusiform, glabrous except for stout pair of medial setae and a minute terminal seta on penultimate labial palpomere, antennae reaching well past base of pronotum, basal antennomeres with one long subterminal seta, third with a ring 4-5 smaller setae, segments IV-XI densely pubescent except for thin, interior and exterior glabrous stripes, segments elongate, length 3.7-3.9X width.

Pronotum distinctly wider than long, median length/width at widest point 1.34-1.50, widest just behind midpoint; piceous, all margins thinly paler rufo-piceous, shiny, slightly duller than head due to more evident microsculpturing and punctulae, lateral margins smoothly curved to hind angles, hind angles slightly more obtuse than right, weakly or not produced, lateral bead slightly widening to apical margin, base unmargined and produced at inner basal impressions; inner basal impressions distinct and linear, outer impressions broadly and weakly impressed, lateral setae at apical third, basal setae set close to comer of hind angle. Proepisternum dull from strong isodiametric sculpticells, prosternal process with a slight medial depression, tip of process strongly margined, elytra broad, l/w 1.47-1.50, depressed, piceous, feebly shiny, microsculpture as in pronotum, epipleura with apical-two thirds paler, rufopiceous; humeri with minute tooth, shoulders prominent, stria 1-6 impessed, though often weaker apically and/or not reaching elytral base, striae smooth or with weak, elongate punctures, stria 7 faintly impressed, a single dorsal puncture on each elytron just behind midline on third interval or touching second stria (Fig. 84). Flight wing, hind tibia with 3-4 spines in posterior median row, tarsomere V glabrous ventrally. Abdomen: Ventrites with strong isodiametric microsculpture, male with two, and female with four setae at apical margin of last visible ventrite (Will 1998). Among the different species *Diplocheila retinens* (Walk.),

Fig. 84. *Diplocheila* sp (After Kushawaha et al. 2015)

D. cordicollis (Laf.), *D. polita* (Fabr.) and *Diplocheila latifrons* Dejean were common predators of Lepidopteran insects in India (Saha et al 1992, Kumar 1997).

2.3.39 Genus: LACHNOTHORAX

Lachnothorax sp.

Small (6.00 mm), ant like head triangular, less pubescent, shiny, not punctuate, frons grooved, labrum small, straight and transverse; clypeus straight, small and transverse, antennae moderately elongate, 1st and 3rd joints less hairy, rest joint strongly hairy; eyes large, lateral and prominent; well developed condyliform neck. Pronotum small, doom shaped, dense pubescent, highly and deeply punctured. Elytra oblong-oval pubescent, striate-punctuate, punctures not deep near the apex, humeral angle rounded, nine striae, presence of short scutellary striole, interval convex, apices obliquely truncate, legs slender, long and hairy (Fig. 85). *Lachnothorax tokkia* Gestro was recorded as a predator of other insect pests of sugarcane in Malaysia (Lim and Pan 1980).

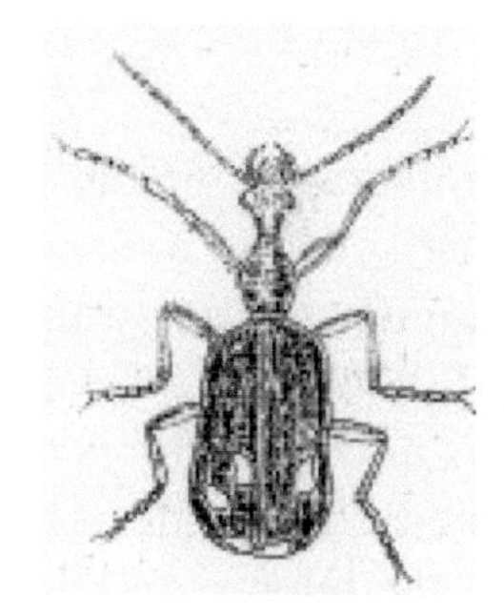

Fig. 85. *Lachnothorax pustulatus* (Dejean) (Images credit: Author)

Lachnothorax biguttata Motschulsky

Species small (6.00 mm) and elongated-ovate. Head, pronotum and elytra black, a pair of rounded orange patches on apical part of elytra; palpi, labrum, clypeus, antennae, legs brown; eyes black. Labrum truncate, transverse, convex and setigerous punctures; clypeus truncate and transverse, fronto-clypeal suture prominent. Head triangular, vertex less hairy and impunctate, antennae 1st to 3rd joint glabrous, only with projecting setae, apex narrower than base, disk closely and deeply punctured. Elytra elongate-ovate, humeral angle rounded, striate-punctuate, punctures coarse, deep and close, not extended to apex, apices obliquely truncate, disk with projecting setae. Leg long and hairy (Saha et al. 1992). *Lachnothorax biguttata* Motschulsky is one of the uncommon predators of rice planthopper in Eastern India.

2.3.40 Genus: SELINA

Selina westermanni Motschulsky

Prothorax subcylindrical, more or less lengthy, head narrowed behind to a condyliform neck, 1st joint of antennae not scapiform, prothorax subcylidrical devoid of lateral borders and whole body pubescent and glabrous, elytra elongate-ovate or nearly parallel-sided, intervals of striae flat or convex (Fig. 86).

Among the different species *Selina westermanni* Mots. was the common predator of sugarcane pests in Malaysia (Lim and Pan 1980).

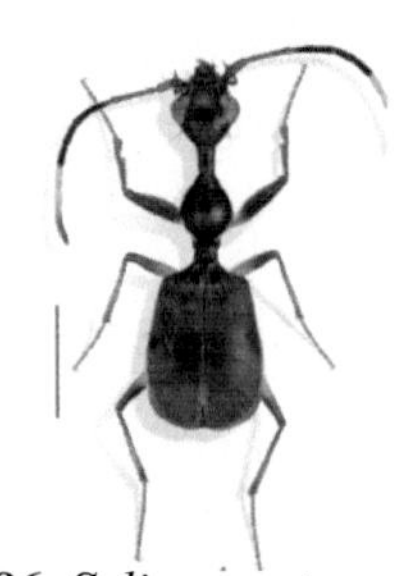

Fig. 86. *Selina westermanni* Motschulsky (Image credit: Erwin T. Zamorano) (From https://upload.wikimedia.org)

2.3.41 Genus: PLANETES

Planetes sp.

Prothorax normally bordered and with sharp edge, antennae joint 1 being scapiform and also antennae slender (Fig. 87) and joint 1 usually about as long as the next three joints together, neck about one-third as wide as the head, tarsal joint 4 not bilobed (Saha et al. 1992). The insect is also a common carabid predator of crop pest in Eastern India.

Among the other species *Planetes pendeleburyl* Andr. was recorded as predator of sugarcane pests in Malaysia (Lim and Pan 1980).

Fig. 87. *Planetes (Planetes) puncticeps* Andrewes (Image credit: Anichtchenko et al. 2007-2016) (From http://carabidae.org/)

2.3.42 Genus: ANTHIA

Anthia sexguttata (Fab.)

The beetles are long and somewhat flattened with a pair of whitish spot on each elytra. Adults measure approximately 4 to 5 cm, black with six relatively large white dorsal spots (four over the elytra and two on the thorax). Other patterns are possible although the pattern is always symmetrical. Males and females are not sexually dimorphic, though it appears males are very slightly smaller than females (Fig. 88).

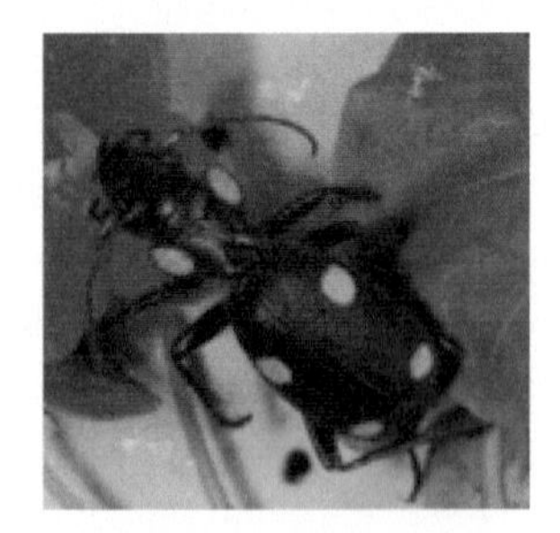

Fig. 88. *Anthia sexguttata* (Fab.) (Image credit: Author) (See Plate 3)

These beetles need a flat ground area to roam and can be quite communal provided there is sufficient food available. They prefer food like small crickets, mealworm, maggots or waxworms in a fairly arid dry surface. Rai et al. (1969) reported that the adult of *Holotrichia consanguinea* Blanch predated upon by this insect. Misra (1975) reported that *Anthia sexguttata* Fab. a new predator of *Pyrausta machaeralis* Walker and *Hyblaea puera* Cramer in India, whereas it is a polyphagous predator in Netherlands (Paarmann 1979). In preference tests, the larvae of *Cyanea bianea, Amsacta mori* (*A. moorei* Butler), *Porthesia scintillans* (Walk.) (*Euproctis scintillans*) and *Dasychira mendosa* Hb. (*Olene mendosa*) were the preferred prey of *Anthia sexguttata* at Regional Research Station, Sekhampur, BCKV, India (Satpathi 2000). The study also showed that *A. sexguttata* preyed on *Diacrisia obliqua* (Walk.) in Eastern India during June and July 1999. Patil and Sathe (2003) reported that *Anthia sexguttata* Horn. is a common predator of *Heiroglyphus banian* (Fab.) in different rice, wheat, jowar and grass fields of India.

2.3.43 Genus: PARENA

***Parena laticincta* Bates**

Body length 7.5 mm, width 1.5 mm, colour brownish with longitudinal groove over the elytra. Elytra partly metallic or not. Head with or without one pair of suborbital setae. Antennomeres 5-11 each with sensory pit on dorsal surface. Mandibles with retinacular ridge of right mandible reduced to retinacular teeth. Lacinia smooth on inner curved. Glossal sclerite with two or three pairs of setae medially, numerous pairs laterally. Mentum with or without one pair of setae. Submentum with two setae laterally each side. Three to four discal setae on each elytron. Males with biseriate adhesive vestiture on fore and mid tarsi. Interval 9 of elytron less than half width of interval 8. Internal sac of male genitalia with or without copulatory piece. Stylomere two of ovipositor with two or more dorsolateral ensiform setae; apex reduced. Reproductive tract with marked elongate tubular sac (Fig. 89).

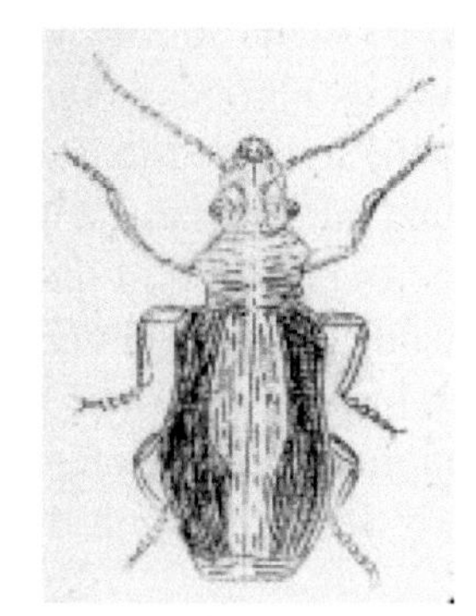

Fig. 89. *Parena laticincta* Bates (Image credit: Author)

The adult of *Parena nigrolineata* Chaud dark brownish, shiny beetle with black border along the lateral and posterior margin of the elytra. The larva as well as the adults of the beetles were predatory in habit, consuming 2 to 3 larvae of the *Nephantis serinopa* Meyr in a week (Mohamed et al. 1982). The beetle of both *Parena laticincta* Bates and *Parena dorsigera* (Schaum) were reported as important predator of the larvae of black headed caterpillar of cocoanut *Nephantis serinopa* Meyr. in India (Rao et al. 1971, Rao 1978, Park et al. 2006, Kumar and Rajagopal 1997). In cocoanut plantations, adults and larvae are also seen attacking the caterpillar *Opisina arenosella* Walker on cocoanut palm trees (Rajagopal and Kumar 1992, Pillai and Nair 1990, Pushpalatha and Veeresh 1995).

2.3.44 Genus: TACHYS

Tachys (Sensu lato*) poecilopterus* (Stein)

The adult of *Tachys* (Sensu lato) *pocciloptera* Bates is piceous or nearly black, larger 2.6 to 3.25 mm. Head with two supraorbital setae on each side, frontal furrows moderately deep and not curving round behind the eyes, mandible with a seta in the scrobe, ligula bisetose, setae placed close together, paraglossae hardly extend beyond the ligula, apical joint of palpi exceeding small and sublate, penultimate joint dilated and pubescent; antennae with the first two joints glabrous; elytra with or without recurrent striole, elytra nine-striate, scutellary stiote as a rule rather slightly impressed, apical recurred striole absent; outer margin of front tibia straight from base to apex.

Among the other species *Tachys sericeus* Motschulsky, *Tachys latus* Peyron, *Tachys truncates* (Nietn.), *Tachys ceylanicus* (Nietn.), *Tachys fumigatus* Motschulsky, *Tachys impressipennis* Motschulsky, *Tachys politus* Motschulsky, *Tachys truncatus* (Nietn.) and *Trachys latus* Peyron are predominant in India (Saha et al. 1992) and *T. sexguttatus* (Fairmaire), *T. brachys* Andrews (Darlington) are common in India and *T. anceps* Leconte and *T. granarius* Dejean in USA (Esau 1968); *T. bisulcatus* in UK (Figs 90 and 91). The species was eastern in its distribution in USA, reaching as far south as Florida, Arkansas, West to Kansas, and north to Ontarico. Lindroth (1966) indicated that the species might sometimes be riparian but has also found in dry sand and in gravelly. Blatchley (1910) indicated that same species occurred in Indiana beneath stones and leaves on damp and wooded hillsides. Dowdy (1947) collected specimens from the leaf layer of a disturbed oak-hickory forest in Missouri. Saraswati (1990) recorded this beetle as an important natural enemy of *Rhopalosiphum nymphaeae* L. infesting *Euryle ferox* Salisb in North Bihar India.

Fig. 90. *Tachys bistriatus* Dufts. (From https://en.wikipedia.org)

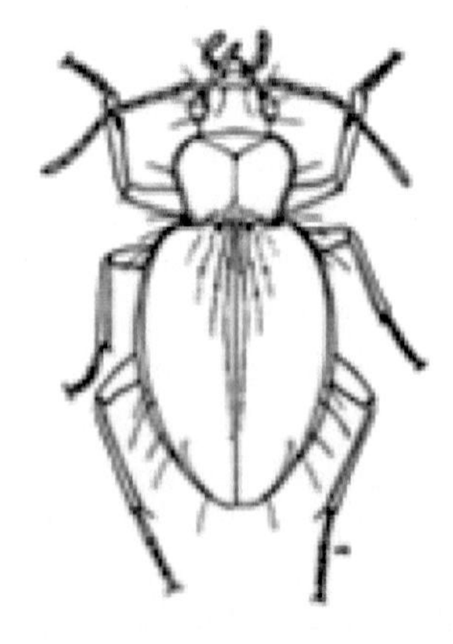

Fig. 91. *Tachys bisulcatus* (After Carl H. Lindroth, *RES* 1974)

2.3.45 Genus: ELAPHROPUS

Elaphropus charis Andr.

Mandible without serration on incisor area. Seta FR9 on front tale more than two times longer than FR5, parietal lateral of seta PA3 with microspines. Dorsal surface with coarsc, isodiametric microsculpture. Elytron with eight micropunctulate interneurs; pronotum longer than wide, constricted at base (Fig. 92). Manjunath et al. (1978a) recorded *Elaphropus charis* Andras the predator of brown planthopper, and other hoppers of rice from India.

Fig. 92. *Elaphropus soort* Mots. (From https://nl.wikipedia.org) (From http://carabidae.org/)

Among the others the beetle of *Elaphropus fumicatus* Motsch was recorded as an effective predator of *Rhopalosiphum nymphaeae* L infesting *Euryale ferox* Salesb in North Bihar, India (Saraswati 1990). The incidence of *Elaphropus latus* (Peyron) was also very common in India (Park et al. 2006).

2.3.46 Genus: BRADYCELLUS

Bradycellus (Tachycellus) anchomenoides (Bates)

Bradycellus (Tachycellus) *anchomenoides* (Bates) is distributed in China, Vietnam, Nepal, Bhutan and Japan. The specimens from Asian continent shows the external characteristics including aedeagi similar to one another. The body 4.2-5.5 mm long (4.4-5.4 mm in examples from China, 4.9-5.5 mm in those from Vietnam, 4.2-4.8 mm in those from Himalaya), the fore and mid tarsi relatively slender, the elytral microsculpture obscure, and the apical lobe of aedeagus short. But the typical Japanese specimens from Honshu, Shikoku and Kyushu are relatively different from them, namely, the body larger in size (5.2-6.0 mm), the pronotum more coarsely punctate in base, the fore and mid tarsi rather strongly expanded, and the elytral microsculpture clearer, and the apical lobe of aedeagus longer (Fig. 93). Such a variation between Japanese specimens and continental specimens were usually found (Ito and Jaeger 2000).

Fig. 93. *Bradycellus harpalinus* (Audinet-Serville) (From https://nl.wikipedia.org)

Bradycellus (Tachycellus) *rupestris* Say was distributed in Long Island, North Carolina, west to Missouri (Casey 1914), Florida, California, Alaska (Leng 1920) and Ontario (Rivard 1964a), where it was found on dry, open ground with moderate to dense vegetation. Webster (1880) found a specimen feeding on a small, white thread-like worm. Among the other species *Bradycellus congener* (LeConte), *Bradycellus rupestris* (Say) were recorded as arthropod predators in cabbage in USA (Schmaedick and Shelton 2000).

2.3.47 Genus: GALERITULA

Galeritula janus Fabricius

Body large about 2 cm long, gaudy. Prothorax bright red, elytra pleasing dark blue with lighter brown legs, antennae and mouthparts. Head slightly wider than long, large prominent eyes; slightly rugose, with broad, poorly developed median carina; posterior half and sides of anterior half with long yellowish, almost erect hairs; two orbital setae. Pronotum longer than wide (length to width ratio: 1.2); as wide as head; anterior margin concave, posterior margin slightly emarginate; widest in the middle; more narrowed anteriorly; sides divergent after the constriction; surface convex, rugose; median sulcus almost erased; covered with long yellowish, backwards directed hairs; two pronotal setae. Scutellum triangular, punctured, with yellow hairs. Elytra twice as wide as pronotum, almost twice as long as wide (length to width ratio: 1.8), widest behind the middle; apex truncate; with nine carinae and two less developed carinulae between every two carinae; a row of deep punctures filling out the carinulae interspace; a row of long yellow hairs (more or less as long as interspace between two carinae) between each carina and carinula; interspaces transversely rugose; scutellar carina usually not joining the first carina (in the holotype the right one joins it). Legs (the holotpe has the right middle femur bifurcated at apex. The body colour is mostly identical to *Galerita orientalis* Schmidt-Goebel (Fig. 94) except head and thorax.

Fig. 94. *Galerita orientalis* Schmidt-Goebel (From https://nl.wikipedia.org)

The insect is distributed in fencerow, Curtiss fencerow, Dekalb Research farm fencerow, San born cornfield and Kalsow rise of USA (Esau 1968). The species is known to occur from Rhode Island, New York State, Maryland, Pennsylvania, Indiana, Iowa and Missouri (Casey 1920) Canada, Kansas and Florida (Leng 1920). This species was frequently taken from underneath the bark of decaying logs and other such damp places in the woods (Adams 1915). This insect was recorded from an Ohio sweet corn fields (Everly 1938), hay fields, alfalfa fields and wasteland areas of Kansas (Walkden and Wilbur

1944). An Ontario record was from dry soil in or near the woods (Rivard 1964a). Dissected specimen of this insect showed that 88% of the food was animal material, mostly caterpillars of various species. Some species collected in an orchard, heavily infested with cankerworms, had 94% animal matter about half of which was cankerworms. Spiders, flies and a moth larva even after completing their food ate cankerworms over and above the normal amount of animal matter (Forbes 1883).

2.3.48 Genus: HARPALUS

Harpalus sp.

Adult beetles dull black, with an elongated oval body and reddish legs. They varied from 1.25–1.6 cm in length and were very mobile. Body relatively slender, shiny black without dorsal microsculpture. First antennomere pale red, second infuscated, 3–11 piceous but slightly paler to apex. Palpi pale red. Legs black with tarsi pale red (pro- and mesotarsi) or piceous (metatarsi), wingless. Anterior margin of labrum almost straight, not arcuate. Anterior margin of epistome very slightly arcuate. Frons with indistinct clypeo-ocular furrows. Eyes moderately protruding, neck normally streched. Mentum without median tooth, sides not thickened, submentum with short pubescence. Pronotum disk flattened. Anterior margin beaded only close to angles, posterior margin entirely beaded. Anterior basis regularly arcuate, angles moderately protruding. Posterior angles obtuse, well-developed. Sides moderately sinuate before hind angles. Basal impression vaguely indicated with some coarse punctures. Sides pubescent, the pubescence extended on the anterior angles and on the posterior basis up to the basal impression. Elytra relatively slender and almost parallel sided, slightly sinuate before apex. Humeral tooth present. Disk flattened, intervals almost flat, striae well marked but impunctate. Scutellar stria with basal setigerous puncture. Normal setigerous puncture in the last third of the third stria or the third interval. Prosternum pubescent, propleura sparsely punctate, metacoxa with few isolated punctures. Last abdominal sternites pubescent with long hairs at each side of median line. Antennae of average size for Harpalus, reaching the beginning of elytra, with pubescence from the 2/3 of the third antennomere. The first antennomere has an extra subterminal seta before the long apical seta. Pro and mesotarsi of males clearly dilated with ventral biseriate adhesive vestiture. All tarsomeres lack the dorsal pubescence existing in Ophonus. Protibiae with 3–4 small spines in the outer distal apex. First metatarsum shorter than second plus third (Serrano and Lencina 2009).

Harpalus rufipes (Degeer) (Fig. 95) consumed more apterous than alate forms of aphids (Loughridge and Luff 1983). The average rate of consumption by individual beetle ranged from 2 to 6 aphids per day. In laboratory, adults consumed up to 130 aphids per day. *H. rufipes* overwintered as both larvae and adults. The overwintered adults became active towards the beginning of May, with their densities peaking by the end of June. Among the other species *Harpalus advolans* fed on termite in India (Kumar and Prasad 1997). *Harpalus affinis* (Schrank) (Fig. 96) was recorded as an omnivorous carabid, feeding mostly on seed, but a smaller amount of animal food was also ingested (Cornic 1973, Sunderland et al. 1995b).

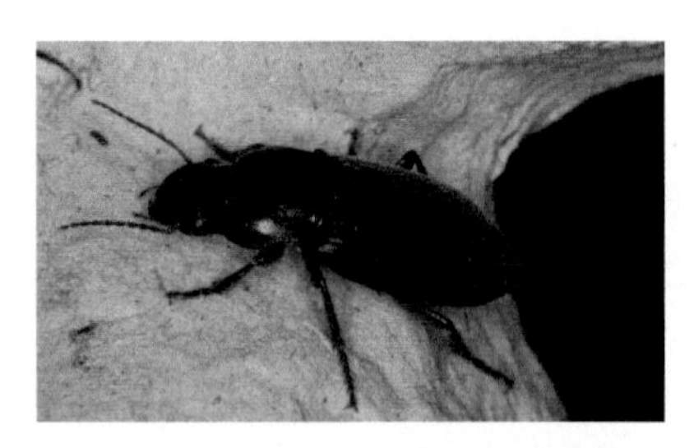

Fig. 95. *Harpalus rufipes* (Degeer) (https://commons.wikimedia.org)

Fig. 96. *Harpalus affinis* (Schrank) (https://en.wikipedia.org/)

2.3.49 Genus: MELAENUS

Melaenus sp.

Elytra slightly more dull in colour than head and thorax, labrum and anterior margin of head shiny; labrum, mandible, labial palpi, maxillary palpi reddish brown. Sides of the head slightly projected in front of eye; a groove runs along lateral margin of the head upto posterior margin of the eye; one supra orbital seta on each side; head and thorax with rugose microsculpture. First four segments of antennae glabrous, round surface and dark reddish brown. Second segment nearly half of the third segment. Fourth segment slightly smaller than 3rd segment. Clypeus convex with one prominent seta on either lateral side. Anteriorly emarginate. Surface moderately punctate with very fine punctures and transverse microsculptures. Labrum anteriorly emarginate with six setae, some small setae on lateral underside. Lateral margin slightly reflexed. Mandible with seta in scrobe. Upper basal region obliquely striate. Hooked at apex; 2nd segment of maxillary palpi glabrous and longer than other segments. Apical segment of labial palpi moderately pubescent (Fig. 97). Mentum short, emarginated anteriorly with one tooth. Lobes and epilobes well developed. One

Fig. 97. *Melaenus piger* (Fab.) (Image credit: Kushwaha, R.K. and V.D. Hegde) (From Zoological Survey of India)

circular fovae with one seta on either side of base of mentum tooth. One seta on either side of base of submentum. One deep puncture on either side of anterior margin of gula. Submentum and gula fused and with transverse striations. Gena deeply punctate. A straight longitudinal sulcus or groove run below eye upto its posterior margin. Prothorax cordate. Single seta on anterior half of either lateral side. Anterior angle obtuse and pointed. Posterior angle slightly diverged outwardly. Abdominal sternite rugose punctated. Apical segment with two small setae suggested female sex. Prosternum longitudinally depressed. Tarsal segments pubescent. On terminal region of elytra an oblique raised area present and interval 1st and 3-8th merge with it. Lateral margin slightly reflexed. Length: 10.5 mm. (Female) (Kushwaha and Hegde 2012). *Melaenus* sp. was one of the occasional predator which fed on rice brown planthopper in India (Kumar & Rajagopal 1997).

2.3.50 Genus: TETRAGONODERUS

***Tetragonoderus quadriguttatus* Dejean**

Elytron with discal setigerous punctures, small; microsculpture mesh pattern isodiametric to distinctly transverse dorsal surface rather shiny or with sericeous lustre; with two spots, preapical one in intervals 5-8 only or unicolorous black. Elytron with mesh pattern distinctly transverse, dorsal surface with more or less sericeous lustre, and two-spotted which is apparently similar to *Tetragonoderus fasciatus* (Haldeman) (Fig. 98). Phallus in dorsal aspect with apical portion short rather broad, apical margin broadly rounded, in lateral aspect projected dorsad. The insect is distributed in South America, from Paraguay northward to Colombia, and the West Indian islands of Grand Cayman, Jamaica, and Hispaniola (http://www.centerfor systematic entomology.org/).

Fig. 98. *Tetragonoderus fasciatus* (Haldeman) (http://www.wikiwand.com)

In general *Tetragonoderus* sp. feeds on termite (Kumar & Rajagopal 1997) and grasshoppers which were damaging to the rice crop (Patil and Sathe 2003) in India.

2.4 Ground Beetles as Common Ant Predator

As a group, ants are important natural predators of many insect pests including flea and fly larvae, caterpillars and termites. However, there are times when it may be necessary to control ants especially when they enter our homes in search of food. Some ant species become problems in lawns and gardens when they build large unsightly mounds or protect aphids, mealybugs, scales and other insect pests from their natural enemies. Ants can also damage plants by tunnelling around the roots causing them to dry out (https//www. planet natural.com). The important carabid predators of ant are recorded as follows.

2.4.1 Genus: CYCLOTRACHELUS

Cyclotrachelus (Evarthrus) sodalis **(Leconte)**

General body length of adult 15 mm, black in colour. Neither sex can fly, for the wing covers were fused shut, and the flight wing underneath vestigial (Fig. 99).

Cyclotrachelus (Evarthrus) sodalis (Le Conte) was reported from Pennsylvania, Ohio, Indiana, Iowa, Nebraska, Missouri, Kansas, Texas and Lousiana (Esau 1968). Dissection of this specimen (Forbes 1880) revealed that none of the food was of vegetable origin. Freitag (1968) reported the species to eat ant.

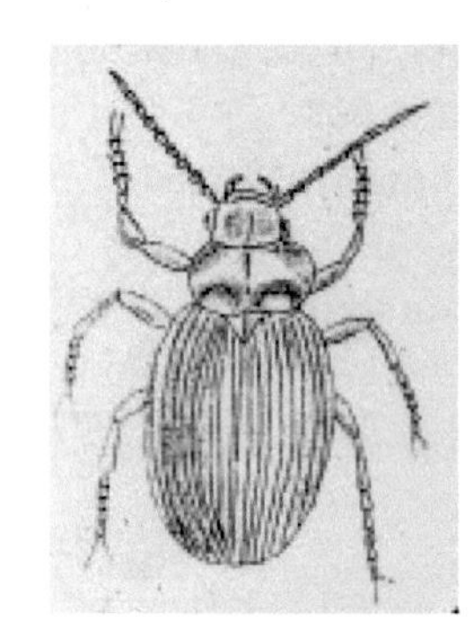

Fig. 99. *Cyclotrachelus sodalis* (Leconte) (Image credit: Author)

2.4.2 Genus: CALOSOMA

Calosoma calidum **Fabricius**

The insect is characterized by a sculpture of elytra of "homodyname" type in which the intervals were of the same height and width, with dotted striae and intervals interrupted by transverse wrinkles. The pronotum is a little restricted to the base with broad rear lobes. The colour of the upper body is greenish brown, on which stood out large foveae with a cupric bottom, in correspondence with the primary intervals but that generally cover at least part of the adjacent ones, length 17 mm (http://www.calosomas.com/).

Burgess and Collins (1917) found that neither the adults nor larvae were effective tree climber. The insect is a potent predator and a voracious feeder on caterpillars and was introduced into Hawaii for the control purpose. Slough (1940) found to eat ant species along with living katydids, stingbugs, cicada nymph, butterfly, chrysalis and measuring worms. Gidaspow (1959) and Ball (1960) differed on adult hunting habits of *C. calidum.* Ball and stated that the adults were known to be incapable of climbing and that they thus confined their hunting activities to the ground. Gidaspow also asserted that the larvae of the species was terrestrial in that they often climbed but that the adults, however, climbed tree easily. Blickenstaff (1965) reported that *Calosoma calidum* Fabricius is one of the few Carabidae recognized as the fiery hunter.

2.4.3 Genus: HARPALUS

***Harpalus caliginosus* Fabricius**

Head distinctly punctate. In females the last tergite (pygidium) demonstrates an angulated apex with abrupt upturn of the margin at the tip (http://bugguide.net/). The habitat of this insect is open, moist ground in general and its distribution ranged from coast to coast and beyond the border (Rivard 1964a). The species was abundant under debris in an Ohio sweetcorn field (Everly 1938) of USA. Slough (1940) concluded from laboratory studies that the insect could take ant pupa but that it probably took more plant materials.

2.5 Acarine Predator

Arthropods serve as prey to an enormous array of predatory animals, ranging from numerous other arthropods to vertebrate species. However, of greatest importance in applied biological control was undoubtedly been various insects and acarine predators, especially Coccinellid and Carabid beetles (Debach and Rosen 1991).

Genus: AMARA

***Amara cupreolata* Putzeys**

Two preapical elytral setae, simple protibial spur, no parascutellar seta, antennae mostly dark (Fig. 100), last abdominal sternum with four apical setae, legs dark, rear setigerous pore of pronotum, slighly away from base (http://bugguide.net/), elytra coppery whereas *Amara ovata* Fabricius (Fig. 100) is black.

Fig. 100. *Amara ovata* Fabricius (From https://en.wikipedia.org)

The greatest number of *Amara cupreolata* Putzeys were distributed at Kalsow Prairie, in the rise and low. Procter (1946) stated their habitat were under logs and stones in damp places. Dambach (1948) recorded this insect in bluegrass field border in Ohio. Forbes (1880) dissected a specimen whose stomach was 90% filled with mite eggs, and also another one with 90% mites and the remainder grass.

2.6 Spider Predator

Spiders which live closer to the ground will suffer from higher rates of predation by ground beetles. The incidences of spider remains in the guts of some species of carabid beetles are isolated and discussed as follows.

2.6.1 Genus: CALOSOMA

***Calosoma calidum* Fabricius**

Calosoma calidum Fabricius is widely distributed from southern Canada to the southern United States. It has also been collected in the district of Columbia and 26 states, including Iowa, Minnesota and Michigan. The range is several provinces in the New England States, south to New Jersey, Pennsylvania, Kentucky, Missouri, Kansas, Colorado and west to Utah (Burgess and Collins 1917). In Canada the insect was known to frequent open, dry fields with low vegetation (Lindroth 1955, 1961) and was also reported by Rivard (1964a) to be found in open ground areas with dry soil. Slough (1940) found this predator to have taken spider in addition to the other insects.

2.6.2 Genus: GALERITULA

***Galeritula janus* Fabricius**

Large about 2 cm long, prothorax bright red, elytra pleasing dark blue. The species is known to Casey from Rhode Island, New York State, Maryland, Pennsylvania, Indiana, Iowa and Missouri, Canada, Kansas and Florida (Leng 1920, Esau 1968) of USA. Adams (1915) stated this species was frequently found from underneath the bark of decaying logs and other such damp places in the wood. The insect was collected from native hay fields, alfalfa fields and wasteland areas in Kansas (Walkden and Wilbur 1944). The stomach of the desected specimen of *G. janus* contained caterpillar of various species along with flies and spiders (Forbes 1983).

2.7 Snail and Slug Predator

Most carabid adults use their well-developed mandibles to kill and fragment prey into pieces. Specialist species attacked snails and seemed to paralyze their prey by biting (Pakarinen 1994) thus prevented the mucus production that was the slug's defense reaction. Many large species ejected a fluid rich in digestive enzymes; subsequently, they consumed the liquid portion of their partially digested prey, sometimes with undigested prey fragments. Larva only consumed extra-orally digested food (Cohen 1995). Slug also appeared to be a significant part of the diet of many generalist carabid predators as well. Early report in the United Kingdom included observation of *Carabus violaceous* which attacked and carried off slugs in its mandible whereas *Harpalus rufipes* Schrak and *Nebria brevicollis* (Fab.) consumed one-day old slug (*Deroceras reticulatum*) in the laboratory (Ayre 2001).

2.7.1 Genus: DICAELUS

***Dicaelus (Paradicaelus) sculptilis* Say**

Elongate oval, broad and robust, dusky black, feebly shining below. Antenae piceous, head less wider, thorax subquadrate, base one-fifth wider than apex, sides feebly curved, the basal third almost parallel, margin flattened and disk uneven, elytra convex, male 17-19 mm length and 7.5 mm width. *Dicaelus sculptilis* Say could have been widely distributed in the preferred habitat but was not able to maintain itself in agricultural lands. The insect was distributed in hilly areas at Cayler Prairie. In addition to Iowa, this species was also known from Ontario, Maryland, Pennsylvania, south to Virginia, west to south Dakota and Colorado, and south to Missouri (Esau 1968). Members of this subgenus were known to eat snails (Ball 1960).

2.7.2 Genus: CARABUS

***Carabus nemoralis* Mueller**

Carabus nemoralis Muller was a beneficial predator as it ate the agricultural pests, slug *Deroceras reticulatum*, in its young stage and also its eggs (Lee and Edwards 1999). It is known that some *Carabus nemoralis* Muller populations will regurgitate foul-smelling brownish-red liquid as a defensive mechanism.

2.8 Weed Predator

2.8.1 Harpalus

***Harpalus affinis* Schrank**

The insect was distributed both in New Zealand (Jorgensen and Lovei 1999) and Denmark (Lovei et al. 2000). Both *Harpalus taradus* (Panzer) and *Harpalus distinguendus* (Dufts.) were also recorded from apple orchards of central Hungary (Horvatovich and Szarukan 1986). Among the other species *Harpalus herbivagus* Say, *H. indigens* Casey, *H. fallax* Le Conte, *H. faunus* Say, *H. longicollis* Le Conte, *H. pennsylvanicus* De Geer, *H. compar* Le Conte, *H. erythropus* Dejean, *H. paratus* Casey, *H. caliginosus* Fab. and *Harpalus affinis* Schrank (Fig. 101) were predominant in USA.

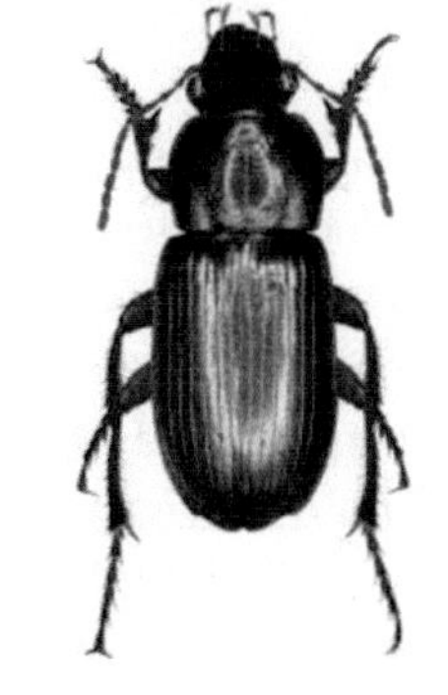

Fig. 101. *Harpalus affinis* Schrank (From https:// commons.wikimedia.org)

H. herbivagus fed on tender grass shoot then later in the season shifted to tender blades and discoloured portions of the grass plants (Webster 1880). The species have been recorded for Iowa and was known from Long Island, Virginia, Nebraska (Casey 1914), and the Pacific Northwest (Hatch 1953). The insect species was collected from not only a wasteland and an old stand of alfalfa, but also from pastureland, roadsides, and a sweetclover field (Walkden and Wilbur 1944) as well as sweet corn fields (Everly 1938).

In Ontario, Rivard (1964a) found this ground beetle common in moist open ground. Casey (1918) indicated that *H. fallax* Le. Conte might be abundant in New Mexico and Colorado and it was further confirmed by the author in an excursion at Sandia Peak, New Maxico, USA during 2016.

Forbes (1880, 1883) reported that this species consumed only vegetable matter, mainly grass tissues and fungi. *H. faunus* Say was widely distributed over eastern North America, ranging from Ontario, to Louisiana, North Dakota, and Arizona. Walkden and Wilbur (1944) recorded this predator from pastures in the Kansas Flint hills. Similar activity periods were reported by Rivard (1964a) from Ontario, where the habitat was moist soil with sparse vegetation. Ball and Anderson (1962) stated that the habitat of the *H. longicollis* Le Conte was damp soil with grassy places near water bodies. *H. pennsylvanicus* De Geer had a wide distribution in the United States, which was collected in every state in the continental United States (Ball and Anderson 1962), as well as the Bahamas, northern Mexico and the southern tier of Canadian provinces. Rivard (1964a) worked in Ontario and collected the species most frequently on open ground in dry soil or commonly in agricultural lands. Herne (1963) found this insect under litter in the vicinity of water.

The species was common under cover in dry pastures of the pacific Northwest (Hatch 1953), under debris in an Ohio sweet corn field (Everly 1938) as well as in wastelands, alfalfa fields and a bluegrass roadside (Walkden and Wilbur 1944). The insect did not consume rootworm larvae offered in the laboratory but it consumed Timothy (*Ambrosia artemisifolia*) seeds and other parts, *Panicum crusgalli* seeds and a nitidulid, *Glischorchilus* (Ips) *fasciatus*. Hendrickson (1930) recorded *H. camper* Le Conte from a prairie near Ames, Iowa, east of the rocky mountains, from Nova Scotia to southern Manitoba and from Florida west to Arizona. Ball and Anderson (1962) listed the species for Iowa. The insect was more common in sweet corn fields (Everly 1938), overgrazed pastures, panic grass and prairie hays (Walkden and Wilbur 1944) and a bluegrass sod fencerow (Dambach 1948). The overall distribution of *H. paratus* Casey was from Michigan to Alberta, then south to Mexico. Several specimens of *H. paratus* Casey were recorded in Texas by Ball (1960) from underneath the litter of oak leaves over dry, sandy, clay soil. *H. caliginosus* Fab. could be represented as the largest *Harpalus* species (Blatchley 1910) and it was widely dispersed in the state of Iowa. Rivard (1964a) reported the habitat in open, moist ground, in general. The author also (1966) found that the insect was common in most Ontario cropland. Wishart et al. (1956) found no evidence for predation by *H. caliginosus* Fab. on cabbage maggot eggs. Webster (1880) observed *H. caliginosus* Fab. to feed on ragweed seeds.

2.8.2 Genus: SELENOPHORUS

***Selenophorus opalinus* LeConte**

Rivard (1964a) found the Ontario habitat in open, dry ground with sparse vegetation (Fig. 102). The insect was distributed in the Curtiss farm fencerow, Kalsow rise, Kalsow and Cayler hill of USA. Blatchley (1910) indicated that the species was frequently collected in Indiana beneath bark. Among the other species *Selenophorus planipennis* LeConte was recorded from Kalsow Prairie and Iowa of USA. Both the predators fed on seeds of weed grasses.

Fig. 102. *Selenophorus parumpunctatus* Dejean (From: https://nl.wikipedia.org)

2.8.3 Genus: DISCODERUS

***Discoderus parallelus* Haldeman**

Hatch (1953) indicated the presence of this species in British Columbia and Washington. The insect was widespread (Fig. 103) across the northern portions of the United States and southern Canada. Walkden and Wilbur (1944) collected it in an overgrazed pasture and in wasteland weed grasses in Kansas of USA.

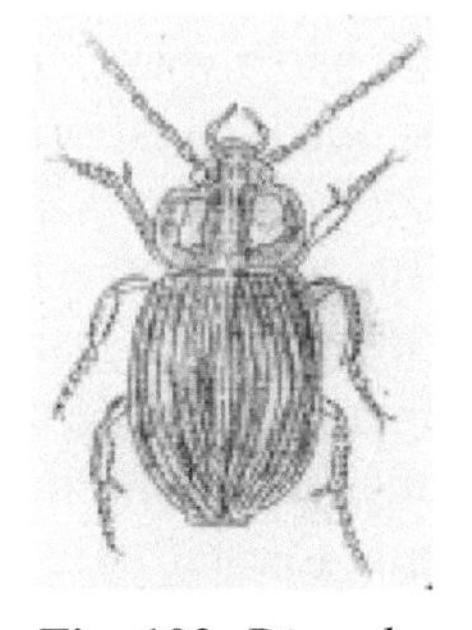

Fig. 103. *Discoderus parallelus* Haldeman (Image credit: Author)

2.8.4 Genus: ANISODACTYLUS

Anisodactylus sp.

Anisodactylus sp fed on grubs and teneral adults of *Phyllophaga anxia* (LeConte) (Coleoptera, Scarabaeidae), a polyphagous pest of lawns, pastures, strawberry, corn and potatoes, Gramineae, soft fruit, cereals, and maize in Quebec, Canada (Poprawski 1994). The ground beetle species was present in most of western and central Europe, and reached as far north as Denmark (Turin et al. 1977), and extended eastward through Russia and China (Fig. 104).

Fig. 104. *Anisodactylus signatus* (Panzer) (From https://en.wikipedia.org)

The species was scarce in Western and Central Europe (Frude et al. 1976), but was frequent in cultivated fields in eastern Europe (Lovei and Sarospataki 1990). *A. signatus* was one of the most common carabids in agricultural field in Hungary (Horvatovich and Szarukan 1986), and occur widely in European

Russia (Berim and Novikov 1983, Metalin 1992), Central Asia (Saipulaeva 1986) and Heilongiiang Province, China (Deng and Li 1981). Species in the genus *Anisodactylus* demonstrated different degrees of mixed feeding. However, both larvae and adults of this species consumed plant materials (Ponomarenko 1969, Hovatovich and Szarukan 1981) mainly germinating seeds. Berim and Novikov (1983) reported that *A. signatus* might be a reluctant predator, preferring plant material to habitat animal prey (Berim and Novikov 1983). Seasonal activities were unevenly documented within this wide range of distribution of this species (Kasandrova and Sharova 1971, Berim and Novikov, 1983). Among the other species *Anisodactylus carbonarius* Say, *A. rusticus* Say, *A. melanopus* Haldeman, *A. ovularis* Casey, *A. harrrisi* Le Conte, *A. agricola* Say, *A. interstitialis* Say and *A. sanctaecrucis* Fabricius were predominant in USA (Esau 1968). *A. carbonarius* Say was distributed in Long Island, New York to Missouri (Casey 1914) Indiana and Iowa (Esau 1968).

The insect was recorded from alfalfa fields, a dropseed wasteland, an overgrazed pasture, a bromegrass pasture, a bluegrass roadside, a prairie hay field, and a wasteland in Kansas (Wilkden and Wilbur 1944). General habitat was open, moist ground with moderate to dense vegetation (Rivard, 1964a). *A. rusticus* Say is common in Indiana, especially in sandy places (Blatchley 1910) and is also found under debris in an Ohio sweet corn field (Everly 1938). The overgrazed pasture, wasteland and fragecrop fields were common habitat of this insect (Walkden and Wilbur 1944). It was observed running on ploughed ground in early spring (Blatchley 1910) hibernating as an adult in winter (Vestal 1913). From the dissected specimens it was observed that 79% of the food was of vegetable origin and fungi (Forbes 1883). *A. melanopus* Haldeman was recorded from Indiana (Blatchley 1910), Pennsylvania, New Jersey and New York State (Esau 1968). Specimens of *A. ovularis* Casey were recorded from the Curtiss fencerow, one in mid-July and the other during early October (Esau 1968). *A. harrrisi* Le Conte was recorded from New foundland to New England, Pennsylvania, Indiana, and west as far as Nebraska (Leng 1920). Open, moist ground with moderate to dense vegetation were the common habitats of this insect (Rivard 1964a). The foods of this insects consisted of only seeds, and parts of grass (Forbes 1883). *A. agricola* Say was collected from the Sanborn fencerow and cornfields of USA (Esau 1968). The incidence of this insect is widespread but it might only be locally abundant. The range was from Iowa, Illinois, Indiana, east to Tennessee, south to Georgia, and west as far as Missouri and Kansas (Blatchley 1910).

A. interstitialis Say was recorded in Long Island, New York, to Missouri (Casey, 1914). The insect was rare and found predominantly in open, dry ground with moderate to dense vegetation and wooden habitat (Rivard 1964a). The insect consumed vegetable matter almost exclusively, only 3% of the food was insect tissue (Forbes 1883). *A. sanctaecrucis* Fabricius was distributed from Rhode Island to Iowa (Casey 1914). The species was found by Vestal (1913) under a board among bunch grass in an Illinois sand Prairie.

A damp, stony place was the habitat of this insect (Procter 1946). The insect consumed June grass seed and plant. Forbes (1883) dissected specimen and found only 14% food of animal origin. The adult overwintering was suggested by the appearance of tenerals during autumn (late September-early October). These overwintered adults probably represented the first, smaller peak of the seasonal activity curves in mid-June, which coincides with the peak of egg production. The second, larger activity peak signaled the emergence of "young" individuals; teneral beetles continued to emerge until early October. The time period between the start of reproductive activity in May and the appearance of teneral beetles in June was too short for these to be the offspring of adults reproducing within the same season; the emerging teneral adult overwintered in the larva or pupal stages. If the thermal developmental threshold of pupae were similar to the thermal threshold of the adult activity in spring, overwintering pupae moulted in to teneral adult earlier. The fact that this did not happen suggested that at least some individuals overwinter as larvae.

The survival pattern of the old beetles varied across regions. The sudden and complete disappearance of old beetles at a particular region and the continued presence of this age class at other regions until October suggested that some adults might die after one year while those in other places might live longer. The extent of overlap between "old" and "young" was also different, with less overlap in some region (Kasandrova and Sharova 1971, Lapshin 1971). This long activity period with the presence of more generations was similar to the seasonal activity of autumn-breeding carabid species like *Harpalus rufipes* (Luff 1980), as opposed to the activity of other spring breeders e.g. *Platynus dorsalis* (Fazekas et al. 1997) or *Clivina fossor* (Desender 1983) which had no surface-active adults after early August.

2.8.5 Genus: STENOLOPHUS

Stenolophus (Agonoleptus) conjuctus Say

Stenolophus conjuctus Say is common in Indiana (Fig. 105), especially in sandy localities (Blatchley 1910), New York pastures and meadows (Wolcott 1937), under debris in an Ohio sweet corn field (Everly 1938), in bluegrass-sodded fencerow and occasionally in shrubby fencerows (Dambach 1948).

Fig. 105. *Stenolophus*sp. (From https://en.wikiwand.com)

Stenolophus (Agonoderus) comma (Fabricius)

Stenolophus (Agonoderus) comma (Fabricius) is a morphological twin (Fig. 106) of *Agonoderus lecontei* Chaudoir (Blickenstaff 1965). Due to the taxonomic confusion and difficulty in distinguishing between these two species the exact relationship between them is not understood (Esau 1968).

The insect is abundant in a sweetcorn field, usually found under debris (Everly 1938), in garden rubbish and also under wet stones (Procter 1946), from a bluegrass sod of fencerow (Dambach 1948) in Ohio. Lindroth (1955) stated the occurrence to be "on more or less moist grassland" and usually in the vicinity of water and even at the pond's edge.

Fig. 106. *Stenolophus* sp. (From https://en.wikiwand.com)

2.8.6 Genus: BADISTER

Badister notatus Holdeman

The range of this insect is mainly the eastern United States, from southern Ontario to Michigan and Iowa; the southern limit is from Virginia to Missouri and far west as Kansas. *Badister notatus* Holdeman (Fig. 107) was recorded predominantly in the prairie, but was able to maintain itself to some degree in certain fencerows (Esau 1968).

Dambach (1948) recorded this insect from a bluegrass sod fencerow. The species was predominantly found on dry, open ground with moderate to dense vegetation (Rivard, 1964a).

Fig. 107. *Badister bullatus* Schrank (From https://en.wikipedia.org)

2.8.7 Genus: MICROLESTES

Microlestes linearis LeConte

Casey (1920) reported *Microlestes linearis* LeConte from Massachusetts, New York, Wisconsin, Iowa, southwestern Utah and Idaho (Fig. 108). Adams (1909) reported a specimen from litter underneath barberry on a rocky ridge of Isale royale.

Dembach (1948) recorded this insect from shruby fencerows and also Osage orange hedgerows in Ohio. Among the other species *M. maurus* Sturm and *M. minutulus* (Goeze) were predominant in USA.

Fig. 108. *Microlestes maurus* Sturm (Gylle.) (https://en.wikipedia.org)

2.8.8 Genus: CALLIDA

Calleida decora Fabricius

Body length 7 mm, head shiny purple, thorax elongated shiny orange with a metallic green elytra. The distribution range given by Leng (1920) is the Gulf States and Florida. Whelan (1936) recorded this insect from grass and shrubs in a Nebraska prairie.

2.8.9 Genus: CYMINDIS

***Cymindis americana* Dejean**

Leng (1920) listed *Cymindis americana* Dejean as occurring only in New Jersey and Indiana (Fig. 109).

The author also recorded *Cymindis americana* Dejean as occurring only in New Jersey and Indiana. Jaques and Redlinger (1946) reported it from Iowa. In Ontario it was found on moist soil with sparse vegetation (Rivard, 1964a).

Fig. 109. *Cymindis humeralis* (Geoffroy In Fourcroy), (From https://upload.wikimedia.org)

2.9 Fungal Predator

Fungi are microorganisms with chlorophyll-less, nucleated, unicellular or multicellular, filamentous body which reproduces by division of its vegetative cells or by formation of various kinds of well defined asexual or sexual spores. These fungal spores are sometime consumed by different insect species, as follows.

2.9.1 Genus: CRATACANTHUS

***Cratacanthus dubius* Palisot de Beauvois**

The insect was recorded from southeastern United States, New Jersey, then Westward to Indiana, Iowa, Arizona (Jaques and Redlinger 1946) of USA. Blatchley (1910) stated that the insect was found in Indiana gardens, the borders of cultivated fields and may be often during unearthed ploughing operations. A specimen was found by Forbes (1880) to have an empty alimentary canal, except for a few fungal spores.

2.9.2 Genus: ANISODACTYLUS

***Anisodactylus signatus* (Panzer)**

From the dissected specimens of *Anisodactylus signatus* (Panzer) it was observed that 79% of the food was of vegetable origin and fungi (Forbes 1883). Among the other species *A. rusticus* Say was common in Indiana (Blatchley 1910) especially in sandy places. Forbes (1883) found the food of vegetable origin and some fungi in the dissected specimens of this insect.

2.9.3 Genus: AMARA

Amara carinata LeConte

The insect was reported from Iowa and west to Montana and Utah, then south to Kansus (Esau 1968) in USA. Forbes (1883) found that all the foods taken by this insect were of vegetable origin, seeds, fungi and other plant tissues. Walkden and Wilbur (1944) collected this predator in Kansas wasteland, sweet clover and alfalfa fields.

2.9.4 Genus: HARPALUS

Harpalus fallax LeConte

Casey (1918) indicated that *Harpalus fallax* Leconte might be abundant in New Mexico and Colorado of USA. Forbes (1880, 1883) reported this carabid beetle could consume only vegetable matter, mainly grass tissue and fungi.

Plate-1

Chlaenius virgulifer (Licini) (Image credit: Author)

Chlaenius velutinus (Dufts.) (Image credit: Author)

Author collecting *Nebria* sp. from a grape vine at Israel during 2004

Chlaenius rufifemoratus MacLeay (Image credit: Author)

Plate-2

Pheropsophus occipitalis (Mac Leay)
(Image credit :Author)

Ophionea indica (Thun.) inside a damaged sugarcane stem . (Image credit: Author)

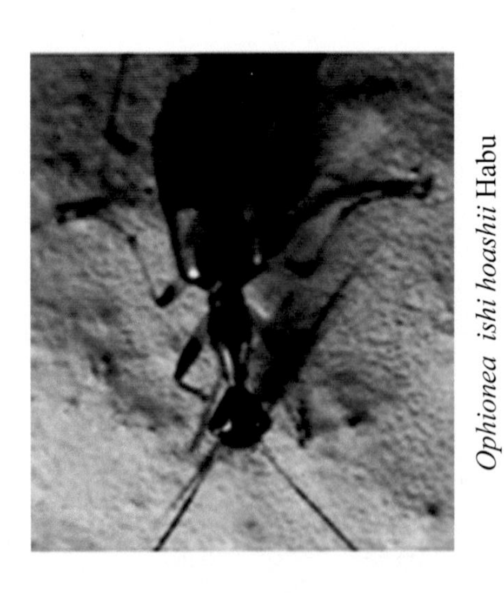

Ophionea ishi hoashii Habu
(Image credit: Author)

Author collecting *Bembidion* sp.
at Taiwan during 2016

Plate-3

Eucolliuris olivieri (Buquet) (Image credit: Author)

Eucolliuris sp. (Image credit: Author)

Drypta japonica (Bates) (Image credit: Author)

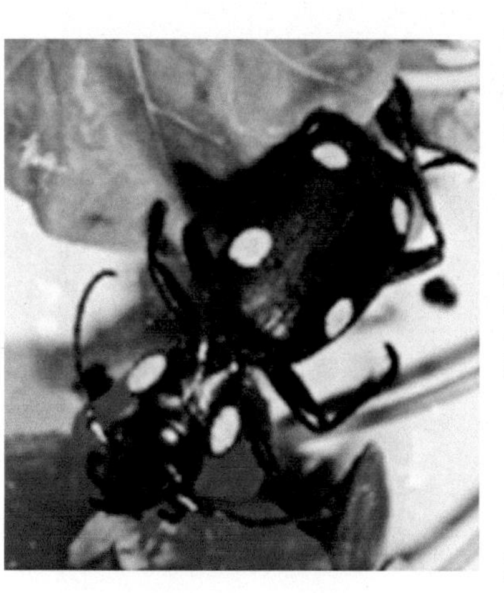

Anthia sexguttata (Fab.) (Image credit : Author)

Plate-4

Author observing ground beetle from a damaged Persimon fruit at Israel during 2004

Author used a special trap to collect ground beetle from an olive garden at Israel during 2004

Author studying biology of ground beetle at the Exosex laboratory in Winchester, United Kingdom during 2011

Author collecting ground beetle in Exosect laboratory at Winchester, United Kingdom during 2011

3

Mass Rearing of Ground Beetle

3.1 Factors Affecting Rearing of Ground Beetle

3.1.1 Predator Rearing Condition

Adults were caught in dry pitfall traps placed in crop fields whereas some adults were found in light trap. Females and males were kept separately in large petri dishes (19 cm diameter) with a layer of sand in the laboratory at natural daylength and temperatures ranging from 20°C to 28°C. The beetles were fed on high-protein frozen (–18°C) larva (*Heliothis armigera* Hub.) for *Harpalus affinis* Le Conte or three times weekly for *Nebria brevicollis* (Fab.). The sand were watered twice a week.

For better rearing efficiency, the feed was also treated with 1% bovine pancreatic inhibitor (BPTI) containing 22.6 µg/g insect (Burgress et al. 2002). During the first five weeks, no beetles were experiencing a remarkable change in body mass irrespective of treatments. After the 5th week a small decrease and then increase was noticed except for the beetles of the BPTI treatment that only lost body mass. After week 8, the beetles of the control treatment were almost constant. The body mass of the beetle in the control BPTI treatment decreased from 51.13 mg to 43.17 mg in week 11 followed by an increase to 48.30 mg in week 12. The BPTI control and BPTI treatments were experiencing small decrease and increase and the BPTI control was decreasing from 45.6 mg to 38.72 mg in week 12. Special devices are prepared in Israel and United Kingdom (see Plate 4) for collection as well as rearing of ground beetles in laboratory.

3.1.2 Preferred Food of Ground Beetle

Blatchley (1910) stated that the female of a pair in copulation was observed feeding upon seeds of ragweed and consumed them at the rate of one every 40 seconds during the 15 minutes of observation. Adams (1915) also reported observing it to feed on ragweed flowers or seeds. Slough (1940) confined *Calosoma calidum* Fab. in to different foods and found them to refuse boiled egg white, dry wheat, and fat meat but to accept cheese, boiled egg yolk, lean meat and banana. The author also found that in the laboratory *Pterostichus lucublandus* Say did not feed on dry wheat but fed on boiled wheat, readily take lean meat and large larvae, and is slightly cannibalistic. The author concluded from laboratory studies that the *Harpalus caliginosus* Fab. can take living animal materials but that usually can take more plant materials. Boiled wheat, peach, cheese, boiled egg yolk, fat, meat, banana, "inch-worm", a tree cricket, and ant pupae were consumed by the adult. During laboratory feeding experiments the author observed that the *Harpalus pennsylvanicus* De Geer accepted inch-worm", lean meat, banana, cheese, boiled eggwhite, boiled egg yolk, and boiled wheat. Shelford (1963) reported that *Amara obesa* Say. utilized bison dung as a food source in addition to grasshopper eggs.

3.1.3 The Effect of Initial Body Mass on Mortality of Ground Beetle

Among the beetles included in the feeding experiments there were remarkable differences in body mass: the smallest weighed 29.5 mg and the largest 70 mg. The possible relationship between size (initial body mass) and survival was examined by regression (Fig. 110).

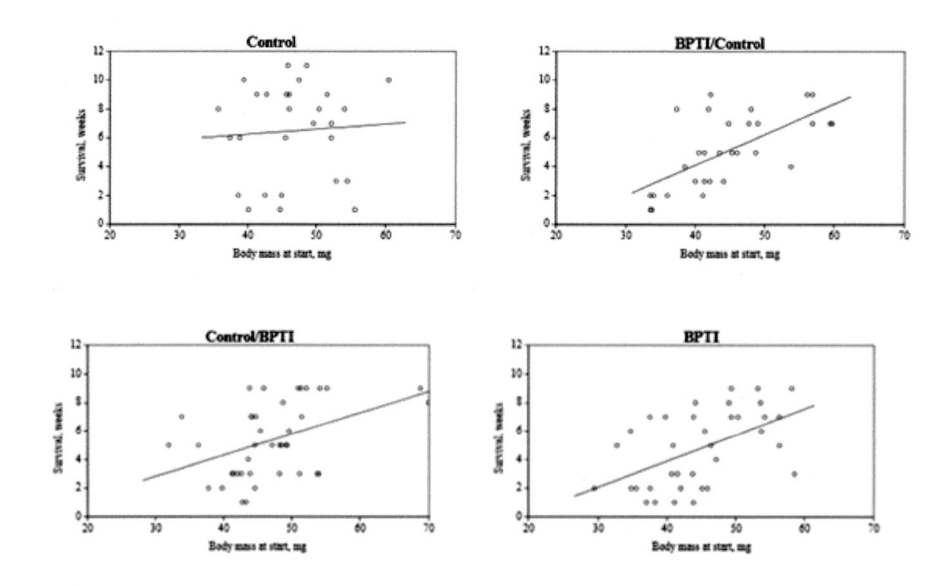

Fig. 110. Survival in relation to initial body mass in adults of *Harpalus affinis* under four different feeding regimes under laboratory conditions. BPTI denotes the proteinase inhibitor (bovine pancreatic trypsin inhibitor). (After Lövei 2008)

There was no significant differences of initial body masses on survival in the control treatment ($r^2 = 0.0043$. $P > 0.710$, but in all other treatments, where beetles consumed 1% bovine pancreatic inhibitor or BPTI-containing prey, larger beetles lived longer i.e. there was a higher mortality of smaller beetles ($P < 0.0001$-0.005).

3.1.4 Effects of 1% Bovine Pancreatic Inhibitor BPTI-fed *Heliothis armigera* (Hub.) on Survival of *Nebria brevicollis* (Fab.)

Prey type had no effect on beetle survival. Both groups declined in numbers over the 24 days of the experiment, with 78% of control and 72% of BPTI-fed prey-fed (AF prey-fed) beetles remaining alive on the final day. The body mass values of control and AF prey-fed beetles did not differ significantly from each other at any weighing occasion. However, there were significant differences in the changes in beetle masses (Fig. 111). On an average, all beetle gained mass during each interval, except for the intervals between day 10 and 14 and day 17 and 21, when both groups lost masses. During those two intervals, the AF conversly, between day 21 and 24 the AF prey-fed beetles gained significantly more masses than control beetles ($P < 0.05$, ANOVA). Covariate analysis of the effect of prey type on body mass change using beetle mass at the beginning of each interval as a covariate showed that beetle size had no influence on the magnitude of mass change observed.

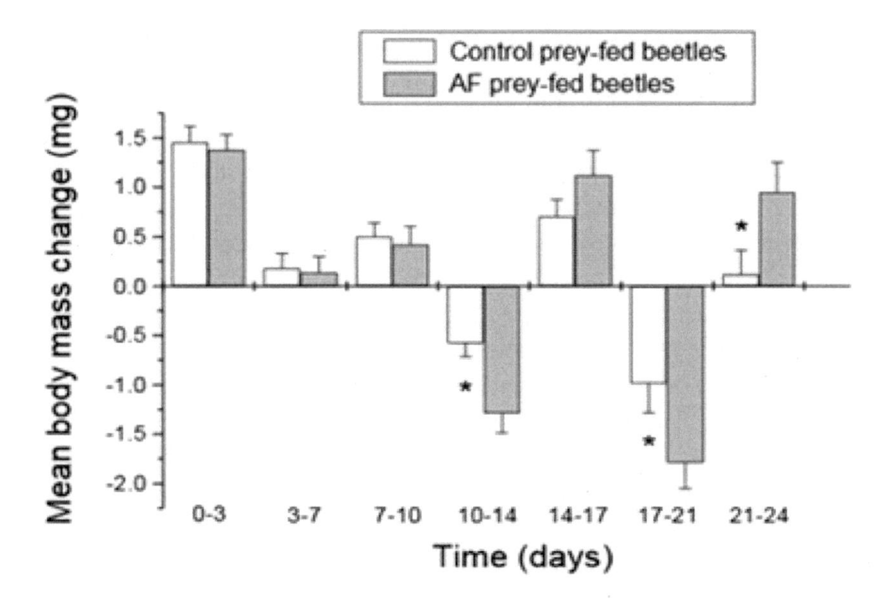

Fig. 111. Mean changes, from one weighing occasion to the next, in the body mass of *Nebria brevicollis* (Fab.) adults supplied with prey (larvae of *Heliothis armigera* (Hub.)) fed with control diet or diet with 0.5% (w: w, fresh mass) BPTI added. Error bars denote the standard error of the mean. Asterisks indicate significant differences between control and BPTI treatments (ANOVA, $p < 0.05$) for each weighing occasion, $N = 97$ for both treatments. (After Lövei 2008)

Table 8. Mean digestive protease activity levels (pmol/min/mg insect) in adult *Nebria brevicollis* (Fab.), values without a letter in common differ significantly from others for the same enzyme type, $P < 0.05$, ANOVA of all beetles. Asterisks indicate significantly higher values in pairwise comparisons of experimental beetles only ($P < 0.05$, ANOVA)

Protease	*Bettles consuming control prey*			*Beetles consuming BPTI-fed prey*			*Field collected beetles*		
	Mean	*s.e.*	*n*	*Mean*	*s.e.*	*n*	*Mean*	*s.e.*	*n*
Chymotrypsin	18.5[a]	1.99	76	38.1[a*]	4.34	69	105.7[a]	26.5	12
Elastase	96[a]	11.2	76	216[ab*]	25.3	69	302[b]	74.7	12
Trypsin	176[a*]	20.4	76	71[a]	8.7	69	632[b]	145	12
LAP	310	19.7	76	331	24.6	69	311	50.8	12
Cystine proteases	18.2[a]	1.65	51	15.8[a]	1.65	40	128[b]	34.0	12

(After Lövei 2008)

The final mass of beetles in the experiment (55.69 mg, s.e. = 0.82 mg, $n = 76$ for controls; 55.48 mg, s.e. = 0.90 mg, $n = 69$ for BPTI prey-fed) did not differ significantly from those of beetles collected from the field at the same time for enzyme analysis (55.66 mg, s.e. = 3.31 mg, $n = 13$) ($P < 0.05$ ANOVA).

3.1.5 Prey Consumption and Bodymass Changes

Equal amounts of control and BPTI-fed prey were supplied to beetles on each weighing day, except for day 7, when the BPTI prey-fed beetle received significantly more prey than the control. Between day 7 and 10, these beetles also consumed significantly more prey than their control counterparts ($P < 0.5$, ANOVA). BPTI prey-fed beetles also ate significantly more than the controls between day 21 and 24 ($P < 0.05$, ANOVA), even though their food supplies were equal. Beetles in both groups were given more prey from day 14 onwards than they have received earlier in the experiment. This was in response to the body mass losses first observed on Day 14 and also because there were more *H. armigera* larvae available by than prey type had no effect on the proportions of food consumed to food supplied (food consumed between Days X and Y/food supplied on day X).

Prey type had a significant effect on the abilities of beetles to convert the food they consumed into bodymass change (interval mass change/food consumed) between days 3 and 7 and days 17 and 21 ($P < 0.05$, ANOVA). During both intervals, beetles converted control prey into bodymass with greater efficiency than BPTI-fed prey. The final mass of beetles in the experiment *(55.69 mg, s.e. = 0.82 mg, $n = 76$ for control; 55.48 mg, s.e. = 0.90 mg, $n = 69$ for BPTI pre-fed) did not differ significantly from those of beetles collected from the field at the same time for enzyme analysis (55.66 mg, s.e. = 3.31 mg, $n = 13$) ($P < 0.05$, ANOVA).

3.1.6 Treated *Nebria brevicollis* (Fab.) and Its Interaction with *Platynus dorsalis* Pont. and *Brachinus exploidens* Dufts

Both *P. dorsalis* (Hokkanen and Holopainen 1986) and *B. explodens* (Kromp 1989) were abundant in biologically managed fields, but rare in conventionally managed ones in Europe. Both species were present but few were captured in the insecticide-treated part of the same apple orchard (Meszaros 1984a). *P. dorsalis* preferred weedy plots in arable land (Powell et al. 1985) especially in spring (Jensen et al. 1989) and the density of the species was positively correlated with weed cover in England (Coombes and Sotherton 1986). The high abundance of *P. dorsalis* in the study area was probably also due to *N. brevicollis* digestive protease activity. Field-collected beetles had significantly higher levels of cysteine protease, chymotrysin and trypsin activities than the beetles used in this experiment (Table 8) ($P < 0.05$, ANOVA). They also had higher levels of elastase than the control beetles, but not the BPTI prey-fed beetles in the experiment.

There were no differences in leucine aminopeptidase (LAP) levels among the three groups of beetles. BPTI prey-fed beetles had significantly higher levels of chymotrysin and elastase, and lower levels of trypsin, than control beetles (Table 8) ($P < 0.05$, ANOVA). LAP and cysteine associated with the dense weedy undergrowth which developed after the abandonment of the orchard. This species in central Hungary demonstrated similar activity and reproductive patterns similar to those in cereal fields in Germany (Kreckwitz 1980). In England, Chiverton and Sotherton (1991) found that an average of 9.7 eggs per female laid a mean of 5.6 eggs over five weeks but this was probably less than the reproductive effort under field condition. These data indicated that *P. dorsalis* females developed one batch of eggs per season but the size of this batch varied considerably, according to local food availability.

Grum's method of calculating total egg output assumes that eggs were laid continuously over the reproduction period and thus might overestimate the reproductive output for *P. dorsalis*. The estimation of total fecundity depended crucially on the eggs laying rates. The study also indicated that this was one of the factors (the other one was egg size difference) which produced the observed large size difference between total reproductive outputs of the two species. The difference meant that while the number of eggs found in individual females was higher in *Brachinus exploidens* Dufts, the total fecundity was similar for two species or possibly even greater for *P. dorsalis*.

Several authors used egg numbers in ovaries as a measure of reproductive output. Grum (1973) estimated the fecundity by assuming a constant oviposition rate throughout the period of female maturity. This method, however, might not be appropriate for all carabid species. Females of *Harpalus aeneus* and *Pterostichus madidus* did not lay more eggs in the laboratory than the peak number observed in the ovaries during the reproductive season.

These species laid a single batch of eggs (Luff 1982). Several species like *Pterostichus versicolor, Calathus melanocephalus* (Linn.) (van Dijk 1979a, b), or *Pterostichus oblongopunctatus* (Heessen 1980) laid eggs countinuously so the total number of eggs laid in a season and the number found in the ovaries were not equal. Grum's method was more suitable for these species, but its application to species that laid a single batch of eggs remained problematic. Reproductive output in ground beetles typically varied among years. It was influenced by food supply and perhaps adult density. Baars & van Dijk (1984) showed a negative correlation between the density of *Calathus melanocephalus* (Linn.) and the number of eggs in their ovaries. In years with low population density, the mean number of eggs laid was 3-4 times higher than the mean number of ripe eggs in the ovaries, whereas in high density year they did not differ (van Dijk 1986). A similar phenomenon was found for *Anisodactylus signatus* in Hungary (Fazekas et al. 1997).

3.1.7 The Impact of GM-laced Prey (PI plant) on Polyphagous Predators

In the short-term experiments, adult *Harpalus affinis* Schrank consumed less of the BPTI-treated prey than control preys. It was not clear what had caused this difference. One obvious possibility was that prey caterpillars carried some food in their guts and the BPTI present in the diet inhibited the predator's feeding. Alternatively, prey caterpillars, due to their diet, could have had a nutritional composition that made them less suitable for the predators. The prolonged effect of the BPTI also seemed counter-intuitive. Lövei et al. (1991) indicated that in several beetle species, a subsequent feeding opportunity hastened the emptying of the earlier meal. Similarly, a fast defecation of the BPTI-treated preys in favour of the later, nontreated one would seem advantageous. This would result in increased feeding of the subsequent, more favourable preys. The cause of this could be due to behaviour (gustatory effect) or a subtle physiological effect.

The BPTI-treated caterpillars affected consumption of *H. affinis* negatively compared with the control caterpillars both in the beetles and in late summer beetles. The difference was most pronounced with the late summer (freshly emerged) beetles. These beetles also responded quite differently from the spring caught beetles with respect to the amount eaten on day 2. Such different food sensitivities by age or sex were not unknown in predatory beetles. Two generations of beetles that overlap, might have different food preferences (Lövei & Sunderland 1996), and could have different metabolism. This was plausible as the two generations feed for different purposes. Young, autumn hatched beetles had fed to build up fat storage so they could survive the hibernation (Wallin 1989) whereas the old generation fed to be able to breed. The result of the first experiment gave a clear indication that seasonality/age of the predators had to be taken into account when considering sensitivity to prey quality. In addition, the result indicates that genetic manipulation of plants could influence interactions at higher trophic levels.

It will be wise to consider these aspects when deciding about the deployment of genetically engineered plants in order to maximize their intended positive effects on agricultural production. Long-term experiments were conducted over 12 weeks. It is not likely that a predator only encountered "transgenic" prey throughout a lifetime like the long-term treatments of the current experiments. However, if transgenic plants get more widespread, the possibility that predators meet 100 prey exposed to GMPs of different traits will increase. Under such conditions, the exposure time increase and long-term experiments such as the current get increasingly realistic. Carabids are sensitive to the quality of their diet. Fecundity in *Calathus melanocephalus* (Linn.) decreased when the beetles are fed aphids (a low-quality food) vs. fruit flies (Bilde and Toft 1999). Not many experiments was conducted on the effects of diet quality on the change of body mass in carabids. Experiments with *N. brevicollis* (Fab.) shows a tendency towards increasing weight loss in short periods of time when fed prey fed PIs (aM-laced plant) but generally no effect of PIs on body mass is found (see earlier). Jørgensen & Lövei (1999) found a significant difference in changes in body masses when feeding *H. affinis* over 48 h. In long-term experiments, no difference was found in mortality and body mass changes. The difference can emerge because adults can adapt to PIs (aM-laced plant) and will only be sensitive during the first few days of exposure.

H. affinis had a mixed diet, eating also seeds and plant materials (Sunderland et al. 1995b), and it is plausible that it had means to overcome an initial PI effect. An omnivorous insect is exposed to the protease inhibitors just as herbivorous insects and it may be the reason why only insects that to some extent are herbivorous (omni- and herbivorous) are able to develop tolerance towards the PI. Many animals are more vulnerable to toxins when they are young than when they are older. It was possible that the beetles were more susceptible when they newly emerged from hibernation (Jørgensen & Lövei 1999). In the long-term experiments, this possibility was tested by feeding *H. affinis* on control prey first and BPTI-prey later in the experiment. No effect of this treatment was found, so *H. affinis* did not seem to be more susceptible to nutritional stress when younger. However, size dependence of life span of predator in the experiments indicated differential sensitivity to nutritional stress. Some predators prefer mobile prey to non-moving ones (Eubanks and Denno 2000). If the prey weakens because of the effect of the PIs, they may become less mobile. The predators may ignore them, eating prey that is more active. This is, however, not the case for *H. affinis*. Preference tests were made in other experiments and no preference to live over dead prey was shown (Jørgensen and Lövei 1999). Many carabids are semelparous and die after one reproductive season (Lövei and Sunderland 1996).

Towards the end of the experiments, mortality of *H. affinis* increased rapidly, probably because of the approaching autumn when many post-reproductive individuals are dead. However, life span in laboratory experiments was a median of >250 days (Lövei and McCambridge 2002), although on a different diet. It was possible that the forced predatory habit caused a different mortality pattern in *H. affinis*. Fecundity in omnivorous carabids could be decreased if only fed animal food (Jørgensen & Toft 1997). The potential tri-trophic effects of a PI-plant will depend not only on the concentration of PI to which the predator is exposed but also on its sensitivity to the PI in question. In this study, beetles consuming prey with a mean BPTI content of 0.0023% (w:w fresh mass) had digestive protease profiles that differed significantly from their control counterparts, suggesting that, even at this low 101 concentration, the PI might have had an effect on the beetle's digestive ability. Reduced trypsin activity in the *N. brevicollis* (F.) beetles that consume BPTI-fed prey suggest that the PI in the prey had a direct impact on this protease, since BPTI is a trypsin inhibitor.

Interestingly, the levels of two other serine proteases, chymotrypsin and elastase, were raised in these insects suggesting some kind of compensatory mechanism. This study revealed the potential for slight and/or transient tri-trophic impacts from PIs on two species of ground beetles. However, beetle survival was not affected by ingestion of BPTI-fed prey and beetle body mass at the end of the experiment was similar to that of field beetles or controls. Mass-related tolerance to nutritional stress were found in both species. More researches are required before the implications for field release of transgenic PI-plants can be ascertained. In particular, further studies in which ground beetles are offered realistic mixtures of prey insect species, including some that feed on PI-expressing transgenic plants, will be valuable.

3.2 Factor Affecting the Mortality of the Different Stages of Ground Beetle

3.2.1 Egg Mortality

Eggs of *Pterostichus oblogopunctatus* Fab. suffered 83% mortality in fresh litter but only 7% in sterilized soil (Heessen 1981). One potential advantage of brood watching can be protection from pathogen, although females had not observed cleaning, surface sterilizing, or even doing anything with their eggs in the egg chamber. However, when abandoned by females, egg quickly became mouldy (Brandmayr & Zetto-Brandmayr 1979).

3.2.2 Larval Mortality

Larval mortality is probably a key factor in overall mortality of ground beetles. Because larvae has weak chitinization and limited mobility, they are sensitive to desiccation, starvation, parasites and diseases. Larvae are also cannibalistic. In laboratory and field studies with surface-active larvae of *Nebria brevicollis* (F.), mortality varied between 25% and 97%, depending on food conditions; parasitism caused up to 25% mortality (Nelemans 1987b, Nelemans et al. 1989). The results of similar study with larvae of *Pterostichus oblogopunctatus* Fab. combined with computer simulations, indicate a cumulative mortality rate of 96% for larvae and pupae (Brunsting et al. 1986). The study concludes that events during larval life were most important for population regulation.

3.2.3 Adult Mortality

Up to 41 % parasitism by nematode and ectoparasitic fungi was found on 14 species of *Bembidion* in Norway (Andersen and Skorping 1991). Nematode infection in insects might cause sterility (Poinar 1975), resulting in obvious fitness effect. The benefit of living in exposed habitats could be freedom from parasites; the cost will be higher risk of predation and/or more frequent catastrophic events, such as flooding (Andersen and Skorping 1991). Different studies indicated that predation is an important mortality factor for adults. Hundreds of vertebrate species prey on carabids (Larocelle 1975a, 1975b,1980). The ecological significance of predation pressure by small mammals was also recorded from North America (Parmenter & MacMahon 1988) and England (Churchfield et al. 1991), where, excluding small mammals resulted in an increase in both species richness and density of carabids.

3.3 Reproduction

Facundity ranged from five to ten eggs per female in species with egg-guarding behaviour to several hundreds per female in species that did not guard eggs (Zetto-Bradmayr 1983). Eggs were laid in one batch, several batches in one season, or over several seasons. As many as 30-40% of individuals in a population reproduced in more than one year (Sota 1987, van Dijk 1972, Vlijm et al. 1968, Cartellieri and Lovei 2003). The dependence of fecundity on age was not well understood. For several species, young females had higher reproductive outputs than older ones (van Dijk 1972), whereas the reverse is true in other species (Burgess 1911, Davies 1987, Gergely and Lovei 1987, Sota 1984). Increased mortality during reproduction might result from ecological rather than physiological factors (Calow 1979), such as exposure of reproducing individuals to higher level of external hazards such as predators or disease. In all carabid species, the variable egg production was related to amounts of food. The first priority of the adults was to meet energy demand for survival and use the surplus for reproduction (see Plate 4). Under condition of limited food supply, this survival—but not reproduce—option enabled predators to survive until better food conditions allowed reproduction

(Mols 1988, Wiedenmann and O'Neill 1990). Data from Europe (van Dijk 1983, van Dijk 1994), Japan (Sota 1984), and North America (Lenski 1984) indicated that carabids in the field regularly experienced food shortage and rarely realized their full reproductive potentials.

In searching for an explanation of carabid fecundity, Grum (1984) found that egg number tended to decrease as body masses increased. Autumn breeders have higher egg numbers than spring breeders, and egg-laying rates were inversely correlated with female mobility (Grum 1984). These results, along with observations of low egg numbers in cave inhabiting species (Deleurance and Deleurance 1964) and of species demonstrating parental care in Europe and New Zealand (Brandmayr and Zetto-Brandmayr 1979), conform to some prediction of the *r* and K-strategy theory. Also, ground beetle species living in unstable habitats had higher egg numbers than relatives which lived under less variable conditions. Similar differences are observed in adult life spans and egg numbers among the Polish and Dutch populations of several species (Grum 1984). However, the *r-K* theory is only one of the hypotheses suggested to explain life-history features.

Appendix A

List of Ground Beetles along with Their Prey Record

1. *Abacetus crenulatus* Dejean: Rice Insect Pest Complex[70]
2. *Abacidus permundus* Say: *Polyphagous[15]
3. *Acupalpus inornatus* Bates: Brown Planthopper[9] and other Planthoppers of Rice[9]
4. *Acupalpus meridionalis* (Linn.): *Polyphagous[39a]
5. *Acupalpus (Egadroma) smargdulus* Fab.: Soybean Pest Complex[18], *Nilaparvata lugens* Stål[9]
6. *Agonum cuprepenne* (Say): Wheat Pest Complex[18]
7. *Agonum dorsalis* (Pont.): *Tetraneura nigriabdominalis* (Saski)[61,14], Polyphagous[5]
8. *Agonum muelleri* (Herbst.): Oilseed and Rape Pests[1], Polyphagous[1a], Soybean pest complex[18], *Sitobion avenae*[29], *Metopolophium dirhodum*[29], *Rhopalosiphum padi*[29]
9. *Agonum quadripunctatum* (Buck.): *Polyphagous [7,29]
10. *Amara aenea* (DeGeer): Grass Seed[22a], *Polyphagous[39a]
11. *Amara anthobia* Villa and Villa: *Polyphagous[39a]
12. *Amara bifrons* Gyllenhal: *Polyphagous[39a]
13. *Amara carinata* (Le Conte): Fungus[16]
14. *Amara consularis* (Dufts.): *Polyphagous[39a]
15. *Amara cupreolata* Putzeys: Mite[16]
16. *Amara eurynota* (Panz.): Oilseed and Rape Pests[1]
17. *Amara familiaris* (Dufts.): Omnivorous[16a], Cutworm[16a], Fly Pupa[16a], Grass Seeds[16a], *Polyphagous[39a]
18. *Amara ingénue* Dufts.: *Polyphagous[39a]
19. *Amara littoralis* Mannerheirn: Soybean Pest Complex[18]
20. *Amara obesa* (Say): Grasshopper's egg mass[18], Corn-Soybean Pest Complex[18]
21. *Amara plebeja* Gyllenhal: Wheat Aphid like *Sitobion avenae*[23], *Metopolophium dirhodum*[22b] and *Rhopalosiphum padi*[22b]
22. *Amara similata* Gyllen.: Predator of Oilseed and Rape pests[1], *Polyphagous[39a]
23. *Anchomenus dorsalis* (Pont.): Oilseed and Rape pests[1], *Polyphagous[1a], Aphid[29,20]

24. *Anisodactylus bionotatus* Fab.: *Polyphagous[1a], Fungus [16]
25. *Aniosodactylus ovularis* (Csaey): Cutworm[16a], Fly Pupa[16a], Grass Seeds[16a], Omnivorous[16a]
26. *Anisodactylus rusticus* (Say): Blue grass[67], Soybean Pest Complex[18]
27. *Anisodactylus sanctaecrucis* (F.): Soybean Pest Complex[18], Omnivorous[16a], Cutworm[16a], Fly Pupa[16a], Grass Seeds[16a]
28. *Anisodactylus signatus* Panzer: *Polyphagous[39a]
29. *Anthia sexguttata* (Fab.): *Holotrichia consanguinea*[47], *Pyrausta machaeralis* Walk.,[40] *Pyrausta machaeralis* Walk.,[40] *Hieroglyphus banian* (Fab.)[42], *Hyblaea puera* Cramer[40], *Cyanea bianea* Wlk.[56a], *Amsacta mori* Butt[56a], *Porthesia scintilans* West[56a], *Dasychira mendosa* (Hub.)[56a]
30. *Asaphidion flavipes* (Linn.): *Macrosiphum avenae*[60]
31. *Badister notatus* Hald.: Pest Complex of Corn-Soybean Cropping System[18], Bluegrass[13,52]
32. *Badister bipustulatus* (Fab.): Oilseed and Rape Pests[1]
33. *Bembidion janthinipennis* (Dejean): Pest Complex of Corn-Soybean Cropping System[18]
34. *Bembidion lampros* (Herbst.): *Polyphagous[39a]
35. *Bembidion mimus* Hayward: Pest Complex of Corn-Soybean Cropping System[18]
36. *Bembidion ovipennis* Le Conte: Polyphagous[18]
37. *Bembidion properans* (Stephens): *Polyphagous[1a]
38. *Bembidion quadrimaculatum* Say: *Macrosiphum avenae* Theo.[60], Pest Complex of Corn-Soybean Cropping System[18]
39. *Bembidion quadripennis* Dejean: Pest Complex of Corn-Soybean Cropping System[18]
40. *Bembidion rapidum* (Le Conte): Pest Complex of Corn-Soybean Cropping System[18]
41. *Bembidion semilunium* Neto.: Hopper pests[13]
42. *Bembidion sobrinum* Boheman: *Nilaparvata lugens* Stål[9], other hopper pests of rice[33]
43. *Bembidion tetracolum* Say: Predator of Oilseed and Rape pests[1]
44. *Brachynus explodens* Dufts.: *Polyphagous[39a]
45. *Brachinus favicollis* Erwin: Pupae of Gyrinidae and Hydrophilidae[51, 56]
46. *Bradycellus (Tachycellus) congener* (Le Conte): Arthropod pests of Cabbage[57]
47. *Broscus cephalotes* (Linn.): *Polyphagous[39a]
48. *Calathus erratus* Sahlberg: *Polyphagous[39a]
49. *Calathus fuscipes* Goeze: *Polyphagous[39a]
50. *Calathus gregarious* (Say): Pest Complex of Corn-Soybean Cropping System[18], Insect[16], Caterpillar[15a]
51. *Calathus melanocephalus* (Linn.): *Polyphagous[39a]
52. *Calathus micropterus* (Dufts.): *Polyphagous[39a]

53. *Calleida decora* (F.): Velvet bean Caterpillar, *Anticarsia gemmatalis* Hub.[19], Larva of *Pseudoplusia includent*[50], Grass[68], *Trichoplusiani* Hübner[69]
54. *Calleida pallipes* Andr.: *Opisina arenosella* Wlk.[28,41,44]
55. *Calleida splendidula* (F.): Caterpillar [41], Black-headed Caterpillar, *Opisina arenosella* Wlk.[28,41,44]
56. *Calosoma calidum* (F.): Katydids, Sting Bug, Cicada Nymph, Ant, Spider, Worm, Larva[59], Pest Complex of Corn-Soybean Cropping System[18], Ant[59]
57. *Calosoma himalayanum* Gerstro: Pest Complex of Corn-Soybean Cropping System[18]
58. *Calosoma maderae var. indicum* Hope.: *Mythimna separate* (Walk.)[48], Termite[48]
59. *Calosoma obsoletum* Say: *Hemileuca olivae* Cock[8] and Grasshopper[8]
60. *Calosoma panagaeoides* Lafarte: Cowpea Aphid[26], *Aphis* spp[26]
61. *Carabus cancellatus* Illiger: Oilseed and Rape pests[1]
62. *Carabus granulatus* Linn.: Aphid[29], *Aphis* sp[29]
63. *Carabus nemoralis* Mueller: * Polyphagous [25, 31,65]; Slug [65], Oilseed and Rape pests[1]
64. *Carabus olympiae* Sella: Snail[29]
65. *Carabus serratus* Say: Pest Complex of Corn-Soybean Cropping System[18]
66. *Carabus violaceus* Linn.: *Polyphagous[39a]
67. *Chlaeniostenus denticulatus elatus* (Erichson): Rice Insect Pests Complex[70]
68. *Chlaenius chlorodius* Dejean: *Hyblea puera* Cramer.[36]
69. *Chlaenius emarginatus* Say: *Lymantria disper* (L.), Pest Complex of Corn-Soybean Cropping System[18]
70. *Chlaenius erythropus* Germar: *Heliothis zea* (Boddle), *Mimestra brassicae* (L.)[32]
71. *Chlaenius micans* (F.): Diamond-back Moth, *Plutella xylostella* (L.)[63]
72. *Chlaenius (Chlaeniellus) nigricornis* Fab.
73. *Chlaenius panagaeoides* (Laferte): Cowpea aphid[58], *Aphis* spp[58]
74. *Chlaenius platyderus* Chaudoir: Pest Complex of Corn-Soybean Cropping System[18]
75. *Chlaenius posticalis* Motschusky: Diamond-back Moth *Plutella xylostella* (L.)[63]
76. *Chlaenius sericeus* (Forster): Pest Complex of Corn-Soybean Cropping System[18]
77. *Chlaenius tomentosus* (Say): Pest Complex of Corn-Soybean Cropping System[18]
78. *Chlaenius tricolor* Dejean: Pest Complex of Corn-Soybean Cropping System[18], Cutworm[16a], Fly Pupa[16a], Grass Seeds[16a], Omnivorous[16a]
79. *Chlaenius velutinus* (Duftschmd): *Cydia pomonella* (L.)[51]

80. *Chlaenius virgulifer* (Licini): *Chilo partellus* (Swin.)[49]
81. *Chlaenius viridis* Men.: *Spodoptera litura* (F.)[64], *Lamprosema* sp[64], *Sylepta derogata*[64]
82. *Craspedophorus angulatus* (Fab.): *Hyblaea puera* Cramer[36]
83. *Cratacanthus dubius* Palisot de Beauvois: Fungus[16]
84. *Cyclotrachelus alternans* (Casey): Pest Complex of Corn-Soybean Cropping System[18]
85. *Cyclotrachelus (Evarthrus) sodalist* LeConte: Ant.[16], Cankerworm[16], Grasshopper Nymphs[59], Leafhopper[59], Stink Bug[59]
86. *Cymindis americana* Dejean: Sparse vegetation[52]
87. *Cymindis pilosus* (Say): Pest Complex of Corn-Soybean Cropping System[18]
88. *Cymindoidea indica* Schmidt-Goebel: Termite[26]
89. *Demetrias atricapillus* (Fab.): Oilseed and Rape pests[1]
90. *Dicaelus (paradicaelus) sculpitilis* Say: Snail[12] and Slug[12]
91. *Discoderus parallelus* Haldeman: Grass[67]
92. *Dioryche nagpurensis* (Bates): *Aphis craccivora* Koch.[4]
93. *Diplocheila retinens* (Walk.): Lepidopteran Insect[26,53]
94. *Diplocheila cordicollis* (Laf.): Lepidopteran Insect[26,53]
95. *Diplocheila latifrons* (Dej.): Lepidopteran Insect[26,53]
96. *Diplocheila polita* (Fabr.): Lepidopteran Insect[26,53]
97. *Dischirius globulosus* (Say): Pest Complex of Corn-Soybean Cropping System[18]
98. *Discoderus parallelus* Haldeman: Pest Complex of Corn-Soybean Cropping System[18], Grass[67]
99. *Dromius melanocephalus* Dejean: Oilseed and Rape pests[1]
100. *Drypta japonica* (Bates): Rice pest[21]
101. *Dychirius globulosus* (Say): Pest Complex of Corn-Soybean Cropping System[18]
102. *Elaphropus anceps* (Le Conte): Pest Complex of Corn-Soybean Cropping System[18]
103. *Elaphropus charis* Andrewes: Brown Planthopper[39b]
104. *Elaphropus fumicatus* Motsch.: *Euryale ferox* Salesh.[55], *Rhopalosiphum nymphaeae* L.[41,55]
105. *Eucolliuris olivier* (Buquet): Rice Pest Complex[21a]
106. *Galerita janus* (F.): Pest Complex of Corn-Soybean Cropping System[18], Caterpillar[16], Cankerworm[16], Spider[16], Flies[16], Moth, Larva[16]
107. *Gnathaphanus licinoides* Hope: Larva of Potato Cutworm[28]
108. *Harpalus advolans*.: Termite[27]
109. *Harpalus affinis* Schrank: Polyphagous pest[1], Animal Food[11,62], Blue Grass[13]
110. *Harpalus calignosus* Fab.: Pest Complex of Corn-Soybean Cropping System[18], Ant[59]
111. *Harpalus distinguendus* (Dufts.): Polyphagous[39a]

112. *Harpalus erraticus* Say: Pest Complex of Corn-Soybean Cropping System[18]
113. *Harpalus fallax* Le Conte.: Fungus[16]
114. *Harpalus faunus* Say: Pest Complex of Corn-Soybean Cropping System[18]
115. *Harpalus herbivagus* Say: Pest Complex of Corn-Soybean Cropping System[18]
116. *Harpalus opacipennis* (Haldemann): Pest Complex of Corn-Soybean Cropping System[18]
117. *Harpalus pensylvanicus* (De Geer): Omnivorous[16a], Cutworm[16a], Fly Pupa[16a], Weed Seeds[16a]
118. *Harpalus rufipes* (Degeer): Aphid[39], Pest Complex of Corn-Soybean Cropping System[18], Polyphagous[19a]
119. *Harpalus tardus* (Panzer): Polyphagous[39a]
120. *Harpalus ventralis* Le Conte.: Pest Complex of Corn-Soybean Cropping System[18]
121. *Lachnothorax tokkia* Gestro.: Pests of Sugarcane[34]
122. *Lebia (Poecilothais)calycophora* Schmdt. Goebel.: Eggs of *Helicoverpa armigera* (Hb.), Colorado Potato Beetle[10]
123. *Lissauchenius venator* (LaFerté): Rice Insect Pests Complex[70]
124. *Loricera pilicornis* Fab.: Oilseed and Rape Pests[1], Polyphagous[1a,37,67], Aphid[6a]
125. *Lymnastis pilosus* Bates: Sugarcane Pests[34]
126. *Macrocheilus impictus* Wied.: Termite[28]
127. *Macrocheilus tripustulatus* (Dejean):
128. *Macrocheilus niger* Andrews: *Pareuchaetes pseudoinsulata*[43]
129. *Melaenus* sp.: Rice Brown Planthopper[28]
130. *Metabletus trancatula* M*ü*ller: *Polyphagous[39a]
131. *Microlestes maurus* (Sturm.): Polyphagous[39a]
132. *Microlestes nigrinus* (Manner.): Polyphagous[1a]
133. *Nebria brevicollis* (Fab.): Polyphagous[1a], Eggs, Larvae of *Helicoverpa armigera* (Hb.)[8a], *Macrosiphum avenae* Theo.[60], Predator of Oilseed and Rape pests[1]
134. *Notiophilus biguttatus* Fab.: Polyphagous[37], Springtail[38], Aphid[38]
135. *Omphra atrata* (Klug.): Ant[45a], Termite[45a]
136. *Omphra complanata* Reiche: Ant[45a], Termite[45a]
137. *Omphra drumonti* Raj, Sabu and Danyang: Termite[45a]
138. *Omphra pilosa* Klug.: Ant [45a,47a], Termite[45a,47a], *Hyblaea puea* Cramer[36]
139. *Omphra hirta* (Fab.): Termite[45a]
140. *Omphra rotundicollis* Chaudoir: Ant[45a], Termite[45a]
141. *Ophionea indica* (Thunb.): Brown Planthopper (*Nephotettix* spp.)[48,54], Rice gall midge *Orseolia oryjae* (Wood-Mason)[25a]
142. *Ophionea interstitialis* (Schmidt-Goebel): Brown Planthopper, *Nephotettix* sp[41] and Green Leafhopper, *Nilaparvata lugens* (Stål.)[41]

143. *Ophionea ishii hoashi* (Habu): Leaf Folder Larvae[21]
144. *Ophonus puncticeps* Stephens: Polyphagous Pests[1a]
145. *Oxylobus dekkanus* Andr.: Termite[26], Polyphagous[48]
146. *Oxylobus punctatosulcantus* Chaudoir: Polyphagous[48]
147. *Panageus cruxmajor* Linn.: *Polyphagous[39a]
148. *Parena drosigera* (Schaum): *Nephantis serinopa* Meyr.[28,41,49], *Opisina arenosella* Walk.[44,46,48]
149. *Parena laticincta* Bates: *Nephantis serinopa* Meyr.[28,41,49], *Opisina arenosella* Walk.[44,46.48]
150. *Pheropsophus* (*Stenaptinus*) *andrewesi* Jedlicka
151. *Pheropsophus aequinoctialis* (Linn.): *Tricoplusia ni* (Hub.)[16a]
152. *Pheropsophus marginatus* (Dejean): Rice Insect Pests Complex[70]
153. *Pheropsophus occipitalis* (Macleay): Eggs of Mole Cricket[17], *Oryctes rhinoceros* Linn[2]
154. *Planetes pedelebyryl* Andr.: Sugarcane Pest Complex[34]
155. *Poecilus chalcites* (Say): Pest Complex of Corn-Soybean Cropping System[18]
156. *Poecilus cupreus* (Linn): Polyphagous[1a,19a], Oilseed and Rape pests[1], aphid[29], *Sitobion avenae* (F.)[29], *Metopolophium dirhodum* (Walk.)[29], *Rhopalosiphum padi* (L.)[29]
157. *Poecilus lucublendus* (Say): Pest Complex of Corn-Soybean Cropping System[18], Cutworm[16a], Fly Pupa[16a], Grass Seeds[16a], Omnivorous[16a]
158. *Poecilus versicolor* (Sturm.): *Polyphagous[39a]
159. *Pterostichus coracinus* (Newman): Pest Complex of Corn-Soybean Cropping System[18]
160. *Pterostichus cupreus* Linn.: Polyphagous[65]
161. *Pterostichus femoralis*: Pest Complex of Corn-Soybean Cropping System[18]
162. *Pterostichus lucublandus* Say: Armiworm[30]
163. *Pterostichus melanarius* (Illiger): *Polyphagous [1a], Pest Complex of Corn-Soybean Cropping System[18], *Aphis fabae* Scop.[22,29,38], Oilseed and Rape pests[1], *Polyphagous[39a], *Rhopalosiphum padi*[66]
164. *Pterostichus niger* (Schaller): Polyphagous[65]
165. *Pterostichus nigrita* (Payk.): Polyphagous[65]
166. *Pterostichus permundus*: Pest Complex of Corn-Soybean Cropping System[18]
167. *Risophilus atricapillus* Linn.: *Aphis fabae* Scop.[12]
168. *Scarites indus* Oliv.: Termite[26]
169. *Scarites* (Orientolobus) *lucidus strigiceps* Quedenfeldt: Rice Insect Pests Complex[70]
170. *Scarites punctum* Wiedem: Insect Predator[21,53]
171. *Scarites subterraneus* F.: Caterpillar and Wireworm[30], Pest Complex of Corn-Soybean Cropping System[18], Cutworm[16a], Fly Pupa[16a], Grass Seeds[16a], Omnivorous[16a]

172. *Selenophorus opalinus* Le Conte: Polyphagous pest[52]
173. *Selina westermanni* Motsch.: Insect Predator of Sugarcane[34]
174. *Stenolophus comma* (F.): Pest Complex of Corn-Soybean Cropping System[18]
175. *Stenelophus (Agonoleptus) conjuctus* Say: Blue Grass soded[13]
176. *Stenelophus (Agonoleptus) comma* (Fab.): Grassland[33]
177. *Stenelophus ochoropezus* (Say.): Omnivorous[16a], Cutworm[16a], Fly Pupa[16a], Weed Seeds[16a]
178. *Syntomus foveatus* (Geoffroy in Fourcroy): *Polyphagous[39a]
179. *Synuchus impuctatus* (Say): Pest Complex of Corn-Soybean Cropping System[18]
180. *Tachys (sensu lato) poecilopterus* (Stein): *Rhopalosiphum nymphae* L.[55]
181. *Tetragonoderus quadriguttatus* Dejean: Termite[28], Grasshopper[42]
182. *Trechus quadristriatus* (Scrank.): *Polyphagous in Grass-Pinebeech[37], Predator of oilseed and rape pests[1], *Polyphagous[39a]

(*Polyphagous/Omnivorous – Feed on live prey, carrion, plant materials including seed, phytophagous)

Literature cited:

1. Alford, 2002
1a. Anjum – Zubair et al., 2015
2. Atwal, 1986
3. Ball, 1960
4. Bhat, 1984
5. Brandmayr and Brandmayr, 1979
6. Briggs, 1957
6a Bryan and Wratten, 1984
7. Buakowski, 1986
8. Burgess and Collins, 1917
8a. Burgess et al., 2002
9. Chiu, 1979
10. Comstock, 1960
11. Cornic, 1973
12. Davies, 1987
13. Dambach, 1948
14. Edward et al., 1979
15. Esau, 1968
15a. Forbes, 1883
16. Forbes, 1880
16a. Frank, 2007
17. Frank et al., 2009
18. French et al., 2004
19. Fuller, 1988
20. Griffith et al., 1985
21. Heinrichs, 1994
21a. Heinrich and Barrion, 2004
22. Hance, 1987
22a. Honeck et al., 2003
23. van Huizen, 1979
24. http: ent.psu.edu/extension
25. Kennedy, 1994
25a. Kobayashi et al., 1995
26. Kumar, 1997
27. Kumar and Rajagopal, 1990
28. Kumar and Rajagopal, 1997
29. Lang, 2003
30. Laub and Luna, 1992
31. Lee and Edwards, 1999
32. Lesiewiez et al., 1982
33. Li et al., 1983
34. Lim and Pan, 1980
35. Lindroth, 1955
36. Loganathan and David, 1999
37. Loreau, 1984c
38. Loreau, 1984

39. Loughridge and Luff, 1993
39a. Lövei, 2008
39b. Manjunath et al., 1978
40. Misra, 1975
41. Park et al., 2006
42. Patil and Sathe, 2003
43. Perera, 1981
44. Pillai and Nair, 1990
45. Poprawski, 1994
45a. Prasad and Rajagopal, 1990
46. Pushpalatha and Veeresh, 1995
47. Rai et al., 1969
47a. Raj et al., 2012
48. Rajagopal and Kumar, 1992
49. Rao, 1978
50. Richman et al., 1980
51. Riddick, 2008
52. Rivard, 1964a
53. Saha et al., 1992
54. Samal and Misra, 1984
55. Sarswati, 1990
56. Saska and Honek, 2004
56a. Satpathi, 2000
57. Schmaedick and Shelton, 2000
58. Shanower and Ranga, 1990
59. Slough, 1940
60. Sopp and Wratten, 1986
61. Sunderland, 1975
62. Sunderland et al., 1995b
63. Suenaga and Hamamura, 1998
64. Swaminathan et al., 2001
65. Symondson and William, 1997
66. Tamaki and Olsen, 1977
67. Walkden and Wilbur, 1944
68. Whelan, 1936
69. Whitcom and Bell, 1964
70. Woin et al., 2005

Appendix B

Worldwide Distribution of Some Important Genus of Ground Beetles

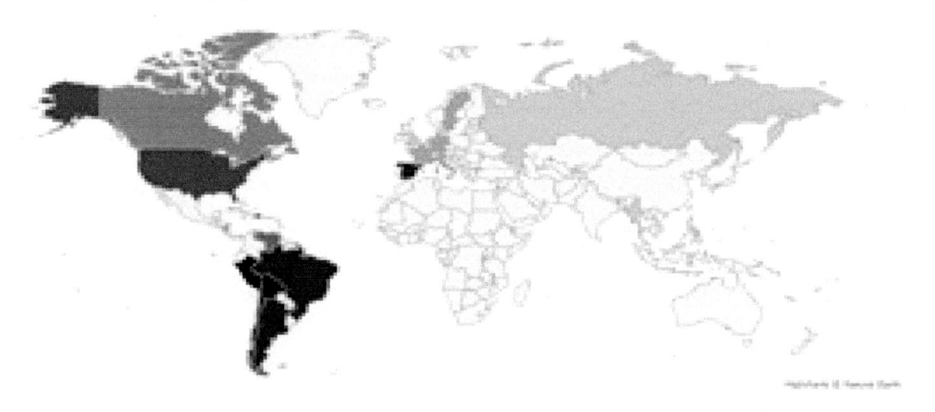

1. *Abacetus* – West Africa (CSIKI 1929), Senegal (Basilewsky 1969, Straneo 1956b), Nigeria, Mauritania (Straneo 1939, 1955), Zaire (Straneo 1956c), India (Saha et al. 1992), Sri Lanka (Bambaradeniya 2016), West Africa (Woin et al. 2005), Vietnam (Park et al. 2006)
2. *Abacidus* – United States of America (Walkden and Wilbur 1944, Esau 1968), Sri Lanka (Bambaradeniya 2016)
3. *Abax* – United Kingdom (Lindroth 1974), Spain (Gutierrez 2004), Europe, Italy, Asia Minor, West Serbia (Curcic and Stankovic 2011)
4. *Acanthoscelis* – South Africa (Hogan 2012)
5. *Acinopus* – Bulgaria (Guéorguiev and Guéorguiev 1995)
6. *Actenipus* – Spain (Gutierrez 2004)
7. *Actenonyx* – New Zealand (Larochelle and Larivière 2013)
8. *Acupalpus* – United Kingdom (Joy 1932, Anderson 1981, Luff 1996a), Taiwan (Chiu 1979), India (Saha et al. 1992), Bulgaria (Guéorguiev and Guéorguiev 1995), United States of America (Larochelle and Larivière 2003, Bousquet 2012), Sri Lanka (Bambaradeniya 2016)
9. *Adelotopus* – New Zealand (Larochelle and Larivière 2013)
10. *Aephnidius* – India (Saha 1992), Arab, Syria, Japan, Australia, Oriental region (Darlington 1970)

11. *Aepus* – France, Canary Islands and the archipelago Madeira in Spain, Albania, UK, Ireland, Norway and Sweden (https://en.wikipedia.org/wiki/Aepus)
12. *Afrotarus* – Israel (Krischenhofer 2010)
13. *Afrozaena* – West Africa from Cameroon to northern Angola (Basilewsky 1968), Senegal (Basilewsky 1965)
14. *Agonocheila* – New Zealand (Larochelle and Larivière 2013)
15. *Agonoderus* – United States of America (Bigger 1934)
16. *Agonotrechus* – Vietnam (Park et al. 2006)
17. *Agonum* – United Kingdom (Dawson 1965, Lindroth 1974, Sunderland 1975, Allen 1977, Edward et al. 1979, Nash 1983, Coombes and Sotherton 1986, den Vries and den Boer 1990, Fowles and Boyee 1992, Duff 1993, Ribera et al. 1996), Netherland (Brandmayr and Brandmayr 1979), India (Saha et al. 1992), Mexico, South America (Bousquet and Larochelle 1993), Germany (Lang 2003), United States of America (Esau 1968, Buakowski 1986, Larochelle and Larivière 2003), Spain (Gutierrez 2004) Vietnam (Park et al. 2006), Iran, Palearctic region (Ghahari et al. 2009), Austria (Anjum-Zubair et al. 2015), Sri Lanka (Bambaradeniya, 2016), Asia Minor, Mt. Caucasus, Iran, Austria, Hungary, France, Germany, Italy, Ukraine, Balkan Peninsula, Serbia Europe, North Africa (Curcic and Stankovic 2011)
18. *Allocinopus* – New Zealand (Larochelle and Larivière 2013)
19. *Amara* – United Kingdom (Fowler 1887, Joy 1932, Allen 1956, Moore 1957b, Lindroth 1974, Speight et al. 1982, Hyman and Parsons 1992, Duff 1992, van Huizen 1977, Anderson et al. 1997), United States of America (Forbes 1880, Esau 1968, Erwin 1979a), Russia (Kryzhanovskij et al. 1995), USA (Texas, California, Oregon, British Colombia) (Hayward 1908), Canada (Poprawski 1994, Esau 1968), Belgium (Loreau 1984), Germany (Lindroth 1955), North Africa, Europe, Bulgaria, Siberia, Armenia, Kazakhastan, Tajikistan,Turkmenistan, Ukraine, Uzbekistan, Transbaikalia, Caucasia, Russia, Iran, China, Japan, Slovakia, Czech Republic, Afghanistan, Crimea, Syria, Turkey (Ghahari et al. 2009), Palaearctic, Afrotropical Region (introduced), North America (introduced), Europe, Mt. Caucasus, Siberia, Central Asia, Central, East and Southeast Europe, Serbia (Curcic and Stankovic 2011), Spain (Gutierrez 2004)
20. *Amarotypus* – New Zealand (Larochelle and Larivière 2013)
21. *Amblystomus* – West Africa, Senegal, Sudan (Basilewsky 1951), India (Saha et al. 1992), Russia (Kryzhanovskij et al. 1995), Bulgaria (Guéorguiev and Guéorguiev 1995), Sri Lanka (Bambaradeniya 2016)
22. *Amblytelus* – Australia, Sydney (Hogan 2012), New Zealand (Larochelle and Larivière 2013)
23. *Amerinus* – United States of America (Esau 1968, Buakowski 1986, Larochelle and Larivière 2003)

24. *Amphasia* – United States of America (Larochelle and Larivière 2003), India (Saha et al. 1992)
25. *Anaulacus* – Sri Lanka (Bambaradeniya 2016), Israel, Vietnam (Park et al. 2006)
26. *Anchista* – Sri Lanka (Bambaradeniya 2016), South Asia, New Guinea (Darlington 1970)
27. *Anchomenidius* – Spain (Gutierrez 2004)
28. *Anchomenus* – United Kingdom, Iran, Central Asia, Europe, Caucasia, Russia, Siberia (Gueorguiev and Gueorguiev 1995, Kryzhanovskij et al. 1995), Turkey (Kesdek and Yildirim 2004, 2010a), Austria (Anjum-Jubair et al. 2015), Germany (Lang 2003, Griffiths et al. 1985), Spain (Gutierrez 2004), Central Asia, Siberia, Europe, Iran, Turkey, Russia, Caucassia, Morocco, Czech Republic, Slovakia, Bulgaria, Maldova, Armenia, Kazakhastan, Turkmenistan, Ujbekistan, Tajikistan (Ghahari et al. 2009), Europe, Asia Minor, Mt. Caucasus, Near East, Central Asia, Siberia, Morocco, Serbia (Curcic and Stankovic 2011), New Zealand (Larochelle and Larivière 2013)
29. *Anisodactylus* – Europe (Frude 1970), India (Saha et al. 1992), Hungary (Horvatovich and Szarukhan 1981), Fazekas et al. 1997), Austria (Anjum-Zubair et al. 2015), Russia, China, Denmark (Turin et al. 1977), USA (Bousquet and Larochelle 1993, Larochelle and Larivière 2003), United Kingdom (Spence and Spence 1988, Hyman and Parsons 1992), Russia (Kryzhanovskij et al. 1995), Spain (Gutierrz 2004), Vietnam (Park et al. 2006), New Zealand (Larochelle and Larivière 2013), Mongoliva, Tuva, Russia, Bhutan, Vietnam, Laos, India, China (Kataev 2015)
30. *Anomophaenus* – New Caledonia (Hogan 2012), India (Adrewes 1929)
31. *Anomotarus* – New Zealand (Larochelle and Larivière 2013)
32. *Anthia* – A West African species recorded from Mauritania (Basilewsky 1970) and Senegal (Basilewsky 1965), Netherland (Paarman 1979), India (Rai et al. 1969, Misra 1975, Saha et al. 1992), Russia (Kryzhanovskij et al. 1995), Iran, Plearctic region (Ghahari et al. 2009)
33. *Anthracus* – East and South Africa (Basilewsky 1951), Tschad (Mateu 1966), Mauritania (Basilewsky 1970), India (Saha et al 1992), Bulgaria (Guéorguiev and Guéorguiev 1995), Sri Lanka (Bambaradeniya 2016)
34. *Antillscaris* – Puerto Rico, El Yunque (Hogan 2012)
35. *Apotomus* – India (Saha et al. 1992), Bulgaria (Wrase, 1992b), Bulgaria, Russia (Guéorguiev and Guéorguiev 1995, Kryzhanovskij et al.1995), Sri Lanka (Bambaradeniya 2016)
36. *Apristus* – Sri Lanka (Bambaradeniya 2016), Mexico, South America (Bousquet and Larochelle 1993), Bulgaria (Guéorguiev and Guéorguiev 1995)
37. *Arame* – Sri Lanka (Bambaradeniya 2016)
38. *Archicarabus* – Spain (Gutierrez 2004), Bulgaria (Guéorguiev and Guéorguiev 1995)

39. *Ardistomis* – Switzerland (Erwin and Halpern 1978)
40. *Argutor* – New Zealand (Larochelle and Larivière 2013)
41. *Aristopus* – West and Central Africa (Basilewsky 1964)
42. *Archicolliuris* – A West and Central African species, recorded from Senegal (Liebke 1931, Basilewsky 1969), Sri Lanka (Bambaradeniya 2016)
43. *Ardistomopsis* – Sri Lanka (Bambaradeniya 2016)
44. *Arrowina* – Sri Lanka (Bambaradeniya 2016)
45. *Asaphidion* – Belgium (Loreau 1984), Denmark (Lindroth 1985), United Kingdom (Speight et al. 1986), Bulgaria (Guéorguiev and Guéorguiev 1995), Iran, Balkan, Peninsula, Bulgaria, Mediterranean countries, Russia, Caucasia (Ghahari et al. 2009), USA (Alaska, Massachusetts), Nepal (Maddison and Ober 2000)
46. *Aspidoglossa* – United Kingdom (Lindroth 1974), Switzerland (Erwin and Halpern 1978), United States of America, Florida (Hogan 2012)
47. *Aulacopodus* – New Zealand (Larochelle and Larivière 2013)
48. *Aulacoryssus* – West and Central Africa (Baehr 2003), Senegal and Gambia (Basilewsky 1950a)
49. *Axinotoma* – Whole of sub-Saharan Africa, except for the southern part (Baehr 2003), Senegal, Gambia (Basilewsky 1950a)
50. *Badister* – United Kingdom (Lindroth 1972, Speight 1976b and 1977, Hansen 1996), United States of America, Europe, Asia Minor, Syria, Israel, Northwest Africa, Serbia (Curcic and Stankovic 2011)
51. *Baenningeria* – Galapagos, Chatham Island (Hogan 2012)
52. *Bembidion*-Senegal (Basilewsky 1950b, 1956), Ranges through the whole of sub-Saharan Africa, Mauritiana (Basilewsky 1964), United Kingdom (Lindroth 1974, Dunning et al. 1975, Turin et al. 1977, Luff 1981b, Speight et al. 1982, Sopp and Wratten 1986), Denmark (Lindroth 1985), Taiwan (Chiu 1979, Li 1983), Sweden (Cheverton 1986), Africa, United States of America (Tamaki and Olson 1977, Larochelle and Larivière 2003), Mexico, South America (Bousquet and Larochelle 1993), India (Saha et al. 1997), North Africa, North America, Europe, Caucasia, Siberia, Russia (Guéorguiev and Guéorguiev 1995, Kryzhanovskij et al. 1995, Neculiseanu and Matalin 2000), Vietnam (Park et al. 2006), Iran, Middle and West Siberia, Russia, Transbaikalia, Ukraine, Turkey, Kazakhastan, Tajikistan (Ghahari et al. 2009), Europe, Asia Minor, West Siberia, North Africa, North America (introduced), Serbia (Curcic and Stankovic 2011), Turkey (Casale and Taglianti 1999, Sahlberg 1912-1913), New Zealand (Larochelle and Larivière 2013), Iran (Najmeh and Hamid 2014), Austria (Anzum-Zubair 2015), Sri Lanka (Bambaradeniya 2016)
53. *Blethisa* – Denmark, Palearctic (Lindroth 1985)
54. *Bodister* – Spain (Gutierrez 2004)
55. *Bothriopterus* – Spain (Gutierrez 2004)

56. *Bountya* – New Zealand (Larochelle and Larivière 2013)
57. *Brachinus* – Northern Africa, Senegal (Basilewsky 1950b, 1969), Indonesia (Krischenhofer 2010), United Kingdom (Speight et al. 1982), India (Saha et al. 1992), Bulgaria, (Guéorguiev and Guéorguiev 1995), Russia (Kryzhanovskij et al. 1995), Spain (Gutierrz 2004), Vietnam (Park et al. 2006), Europe, Asia Minor, Mt. Caucasus, West and Central Asia, Siberia, Baikal area, Serbia, Balkan Peninsula (Curcic and Stankovic 2011), Iran (Najmeh and Hamid 2014), Sri Lanka (Bambaradeniya 2016)
58. *Brachychila* – Malaysia (Krischenhofer 2010)
59. *Brachyodes* – West Africa, Senegal (Lecordier and Girard 1990), Sri Lanka (Bambaradeniya 2016)
60. *Badister* – Ontario, Michigan, Iowa, Virginia, Missouri, Kansus of USA (Rivard 1964a, Esau 1968)
61. *Brachystylus* – USA (Bousquet and Larochelle 1993)
62. *Bradybaenus* – Widely distributed through sub-Saharan West and Central Africa (Basilewsky1950a), Mauritania (Basilewsky 1964), Senegal (Basilewsky 1951), Sri Lanka (Bambaradeniya 2016), Africa, India (Saha et al. 1992, Kumar and Rajagopal 1997)
63. *Bradycellus* – United Kingdom (Fowler 1887, Sharp 1913, Lindroth 1972, Nash 1979, Anderson et al. 1997), North Carolina, Missouri of USA (Casey 1914), Florida, California, Alaska of USA (Leng 1920), Ontario (Rivard 1964a), Bulgaria (Guéorguiev and Guéorguiev 1995),Vietnam (Park et al. 2006)
64. *Bradytus* – Spain (Gutierrez 2004)
65. *Broscus* – Denmark (Lindroth 1985), Bulgaria, Russia (Guéorguiev and Guéorguiev 1995, Kryzhanovskij et al. 1995); Iran, Mediterranean countries (Ghahari et al 2009), United Kingdom (Hogan 2012)
66. *Brullea* – New Zealand (Larochelle and Larivière 2013)
67. *Caelostomus* – West Africa, Senegal (Balilewsky 1964, 1969, CSIKI 1931), India (Saha et al. 1992)
68. *Calathosoma* – New Zealand (Larochelle and Larivière 2013)
69. *Calathus* – United Kingdom (Anderson and Luff 1994), United States of America (Larochelle and Larivière 2003), India (Saha et al. 1992), Central Asia, Europe, Siberia, Iran, Russia, Syria (Guéorguiev and Guéorguiev 1995, Kryzhanovskij et al. 1995); Bulgaria (Guéorguiev and Guéorguiev 1995), Caucasia, Armenia, Crimea, Ukraine (Kryzhanovskij et al. 1995); North Africa, West Asia, Europe, Czech Republic, Slovakia (Hurka 1996),Turkey (Ganglbauer 1905; Sahlberg 1912-1913, Yücel and Sahin 1988, Kesdek and Yildirim 2004, 2010a), Spain (Gutierrez 2004), Europe, Asia Minor, Mt. Caucasus, Siberia, Central Asia, North Africa, Serbia (Curcic and Stankovic 2011), Europe, Mediterranean Country, Palearctic Region, Iran (Ghahari et al. 2009)
70. *Calleida* – Florida, USA (Leng 1920, Mc Carty et al. 1980, Mc. Whoter

et al. 1984, Fuller 1988), Russia (Kryzhanovskij et al. 1995), India (Pillai and Nair 1990, Kumar and Rajagopal 1997, Park et al. 2006), Sri Lanka (Bambaradeniya 2016), Vietnam (Park et al. 2006)

71. *Callistoides* – Distributed through almost the whole of Africa, Senegal (Basilewsky 1969)
72. *Callistomimus* – United States of America, Sri Lanka, India (Saha et al. 1992)
73. *Callistus* – United Kingdom (Lindroth 1974), Bulgaria (Guéorguiev and Guéorguiev 1995), Russia (Kryzhanovskij et al. 1995)
74. *Callytron* – Sri Lanka (Bambaradeniya 2016)
75. *Calosoma* – USA (Slough 1940, Forbes 1983, Larochelle and Larivière 2003), Bulgaria (Guéorguiev and Guéorguiev 1995), Sub-Saharan Africa and Senegal (Baehr 2003), India (Rajagopal and Kumar 1992, Kumar 1997), United Kingdom (Speight et al. 1982), Iran, Maldova, Bulgaria, Rominia, Ukraine, Russia, Mongolia, Armenia, Lithuania, Crimea, Azerbaijan, Daghestan, Kazakhstan, Siberia, China, Caucasia, Czech Republic, Slovakia, Jawa, Turkey, Siberia (Ghahari et al. 2009), Europe, Asia, Mt. Caucasus, Siberia, North Africa, North America, Asia Minor, Iran, Japan (Curcic and Stankovic 2011)
76. *Carabus* – United Kingdom (Lindroth 1974, Turin et al. 1977, Houston 1981), Denmark (Lindroth, 1985), Bulgaria (Guéorguiev and Guéorguiev 1995), Iran (Najmeh and Hamid 2014), United States of America (Larochelle and Larivière 2003), Germany (Lang 2003), Spain (Gutierrez 2004), Iran, Turkey, Russia, Caucasia, Maldova, Ukraine, Kazakhstan, Turkmenistan, Uzbekistan, Tajikistan (Ghahari et al. 2009), Vietnam (Park et al. 2006), Balkan Peninsula, Italy, Hungary, Slovakia, France, Bulgaria, Czech Republic, Austria, Asia Minor, Mt. Caucasus, Armen (Curcic and Stankovic 2011), New Zealand (Larochelle and Larivière 2013)
77. *Carterus* – Bulgaria (Guéorguiev and Guéorguiev 1995), Ukraine, Russia, Azerbaizan, Afghanistan (Kataev 2015)
78. *Casnoidea* – India (Rajagopal and Kumar 1992, Ullah and Jahan 2004), Vietnam (Park et al. 2006)
79. *Cerabilia* – New Zealand (Larochelle and Larivière 2013)
80. *Chlaenius* – A West African species, recorded from Senegal (CSIKI 1931, Basilewsky 1950b 1956, 1961, 1961a, 1964, 1965, 1969), Mouritania (Basilewsky 1970), Zaire (Basilewsky 1953b), United Kingdom (Fowler 1887, Speight 1977, Duff 1993), Iran, United States of America (Larochelle and Larivière 2003), India (Saha et al. 1992), Mexico, South America (Bousquet and Larochelle 1993), South Asia, Australia (Darlington 1970), Bulgaria (Guéorguiev and Guéorguiev 1995), Turkey, Caucesia, Transcaucesia, Iran, Russia, Kazakhstan, Turkmenistan, Uzbekistan, Tajikistan (Ghahari et al. 2009), Vietnam (Park et al. 2006), Europe, Asia Minor, Kazakhstan, North Africa,

Uzbekistan, Turkmenistan, Siberia, Russian Far East, Balkan Peninsula, Mt. Caucasus, the Transcaucasian area, Iran, Israel, Central Asia, Serbia (Curcic and Stankovic 2011)

81. *Chlaeniostenus* – West Africa (Woin et al. 2005), Vietnam (Park et al. 2006)
82. *Chrysocarabus* – Spain (Gutierrez 2004)
83. *Chydaeus* – Laos (Kataev 2015)
84. *Clivina* – West Africa, Senegal (Basilewsky 1964, 1965, 1968, 1969), Mexico, Philippines (Darlington 1970), Switzerland (Erwin and Halpern 1978), Denmark (Lindroth 1985), India (Saha et al. 1992), North America, Central and South Europe, Caucasia, Kazakhastan, Russia, Asia Minor, Turkistan, China, Bulgaria, Siberia (Guéorguiev and Guéorguiev 1995, Kryzhanovskij et al. 1995), Turkey (Casale and Taglianti 1999), USA (Larochelle and Larivière 2003), Spain (Gutierrez 2004), Vietnam (Park et al. 2006), Middle East, Europe, Turkey, Czech Repubic, Slovakia, Caucasia, Russia (Ghahari et al. 2009), United Kingdom (Luff 1887, Anderson et al. 1997, Hogan 2012), New Zealand (Larochelle and Larivière 2013)
85. *Cnemalobus* – Argentina (Hogan 2012)
86. *Cnides* – United States of America (Maddison and Ober 2000)
87. *Coleolissus* – Sri Lanka (Bambaradeniya 2016)
88. *Colfax* – India (Raj et al. 2012), Switzerland (Erwin and Halpern 1978)
89. *Colliuris* – United Kingdom (Lindroth 1961-69), USA (Bousquet 2010)
90. *Colpodes* – Sri Lanka (Bambaradeniya 2016), India (Saha et al. 1992), Philippines, New Guinea (Darlington 1970)
91. *Coptodera* – Malaysia, Perak (Krischenhofer 2010), Sri Lanka, India, Burma, Sarwak (Andrewes 1929), USA (Gamboa and Ortuño 2015)
92. *Coptolobus* – Sri Lanka (Bambaradeniya 2016), USA (Hogan 2012), India (Saha et al. 1992)
93. *Corintascaris* – Africa (Hogan 2012), Sri Lanka (Bambaradeniya 2016)
94. *Coryza* – Sri Lanka (Bambaradeniya 2016), India (Andrewes 1929), Switzerland (Erwin and Halpern 1978)
95. *Cosmodiscus* – India (Saha et al. 1992, Andrewes 1929)
96. *Costitachys* – United States of America, California, Arizona (Maddison and Ober 2000), Russia (Kryzhanovskij 1995)
97. *Craspedophorus* – West Africa, Senegal (CSIKI 1931, Basilewsky 1969), India (Saha et al. 1992, Loganathan and David 1999), Sri Lanka (Bambaradeniya 2016)
98. *Cratacanthus* – Mexico, South America (Bousquet and Larochelle 1993)
99. *Cratohaera* – West Africa (Wiesner 2001), Gambia, Africa (Baehr 2003)
100. *Creagris* – Sri Lanka, India (Raj et al. 2012)
101. *Crepidopterus* – Switzerland (Erwin and Halpern 1978), Madagascar, Africa,Tananariva, (Hogan 2012), Sri Lanka (Bambaradeniya 2016)
102. *Ctenognathus* – New Zealand (Larochelle and Larivière 2013)

103. *Cryptoscaphus* – Malawi, Tanga Province, Africa (Hogan 2012), United Kingdom (Lindroth 1974)
104. *Cychrus* – United Kingdom (Lindroth 1974), Denmak (Lindroth 1985), Balkan Peninsula, Moldova, Ukraine, South Russia (Kryzhanovskij et al. 1995, Curcic and Stancovic 2011), Bulgaria (Guéorguiev and Guéorguiev 1995)
105. *Cyclosomus* – Sri Lanka (Bambaradeniya 2016), Vietnam, China, Thailand (Park et al. 2006)
106. *Cyclotrachelus* – USA (Forbes 1880, Esau 1968, Freitag 1969), Mexico, South America (Bousquet and Larochelle 1993, Bousquet et al. 1993, Larochelle and Larivière 2003)
107. *Cylindera* – Central Africa (Schacht 2000), Senegal (Basilewsky 1969), Iran, Sri Lanka (Bambaradeniya, 2016), Iran, Paleactic Region, Transcaucasia (Ghahari et al. 2009), Europe, East Palaearctic (Curcic and Stankovic 2011)
108. *Cymbionotum* – Northern part of sub-Saharan Africa, Mauritania, Tschad and adjacent countries (Basilewsky 1961a), Russia (Kryzhanovskij et al. 1995), USA (Erwin 2008)
109. *Cymindis* – United Kingdom (Fowler 1887, Williams 1984, Hyman and Parsons 1992, Duff 1993), New Jersey, Indiana of USA (Leng 1920), Mexico, South America (Bousquet and Larochelle 1993), Bulgaria (Guéorguiev and Guéorguiev 1995), Russia (Kryzhanovskij et al. 1995), Spain (Gutierrez 2004), Afghanistan, Armenia, Azerbaijan, Cyprus, Egypt, Georgia, Iran, Iraq, Kazakhastan, Kyrgyzstan, Lebanon, Pakistan, Russia, Soudi Arab,Tadzhikistan, Turmenistan, Yemen (Öncüer 1991, Löbl and Smetana 2003, Kesdek and Yildirim 2007b, Ghahari and Kesdek 2012), Turkey (Öncüer 1991, Kocatepe and Mergene 2004, Kesdek and Yildirim 2007b), Iran (Najmeh and Hamid 2014)
110. *Cymindoidea* – Sri Lanka (Bambaradeniya 2016), Africa, South India (Kumar 1997)
111. *Daptus* – Bulgaria (Guéorguiev and Guéorguiev 1995), Afghanisthan (Kataev 2015)
112. *Deltomerus* – Bulgaria (Guéorguiev and Guéorguiev 1995), Russia (Kryzhanovskij et al.1995)
113. *Demetrias* – United Kingdom (Hyman and Parsons 1992), Russia (Kryzhanovskij et al. 1995), Vietnam (Park et al. 2006), Europe, Asia Minor, Syria, Israel, Northwest Africa, Serbia (Curcic and Stankovic 2011)
114. *Demetrida* – Australia, New Zealand (Larochelle and Larivière 2013)
115. *Derocrania* – Sri Lanka (Bambaradeniya 2016), India (Andrews 1929)
116. *Desera* – Sri Lanka (Bambaradeniya, 2016)
117. *Diachromus* – United Kingdom (Lindroth 1974), Iran, West Asia, Turkey, Caucasia, Turkmenistan (Ghahari et al. 2009)
118. *Diacheila* – Mongolia, Norway, Russia (Amur, Buryat Republic, Chita

Area, Chukotka, East Sayan, Irkutsk Area, Kamchatka, Khabarovsk, Khamar-Daban, Krasnoyarsk, Magadan, N Ural mts., Yakutiya), United States (Alaska) (http://carabidae.org/taxa/polita-faldermann)

119. *Dicaelindus* – Sri Lanka (Bambaradeniya 2016)
120. *Dicaelus* – Ontario, Maryland, Pennsylvania, Virginia, Dakota, Colorado, Missouri of USA (Esau 1968, Larochelle and Larivière 2003)
121. *Diceromerus* – Sri Lanka (Bambaradeniya 2016), India, Hongkong, Laos, Burma, Singapore, Kualampur, Japan, Sumatra (Saha et al. 1986)
122. *Dicheirotrichus* – United Kingdom (Lindroth 1974), Bulgaria (Guéorguiev and Guéorguiev 1995), Georgia (Kataev 2015)
123. *Dicrochile* – Ausrtralia, New Zealand (Larochelle and Larivière 2013)
124. *Diglymma* – New Zealand (Larochelle and Larivière 2013)
125. *Dinoscaris* – Switzerland (Erwin and Halpern 1978), Madagascar, Africa, United States of America (Hogan 2012)
126. *Dioryche* – India (Bhat 1984, Saha et al. 1992), Sri Lanka (Bambaradeniya 2016), Russia (Kataev 2012)
127. *Diplocheila* – India (Saha et al. 1992, Kumar 1997), Sri Lanka (Bambaradeniya 2016), USA (http://bugguide.net)
128. *Discoderus* – United Kingdom (Lindroth 1974)
129. *Distichus* – West and Central Africa, Senegal (Basilewsky 1956, 1968,1969), Spain, Algaida (Hogan 2012), Iran, Nepal, SriLanka (Bambaradeniya 2016), Iran, Caucasia, Mediterranean countries (Ghahari et al. 2009), Switzerland (Erwin and Halpern 1978)
130. *Ditomus* – Bulgaria, (Guéorguiev and Guéorguiev 1995), Russia (Kryzhanovskij et al. 1995), Central Asia, South Europe, Caucasia, Turkey, Syria (Ghahari et al. 2009)
131. *Dixus* – Bulgaria (Guéorguiev and Guéorguiev 1995)
132. *Dolichoctis* – Sri Lanka, Malayasia (Krischenhofer 2010)
133. *Dromius* – United Kingdom (Fowler 1887, Luff 1966, Lindroth 1986, Duff 1993), India (Saha et al. 1992), Russia (Kryzhanovskij et al. 1995), Spain (Gutierrz 2004), Vietnam (Park et al. 2006), New Zealand (Larochelle and Larivière 2013), Sri Lanka (Bambaradeniya 2016)
134. *Drypta* – Widely distributed through almost the whole of sub-Saharan Africa including Senegal (Basilewsky 1968, 1969), Philippines (Heinrich et al. 1990), India (Saha et al. 1992), United Kingdom (Hyman and Parson 1992), Bulgaria (Guéorguiev and Guéorguiev 1995), Russia (Kryzhanovskij et al. 1995),Vietnam (Park et al. 2006), Europe, Asia Minor, Southwest Asia, Turkmenistan, Northeast and South Africa, Serbia (Curcic and Stankovic 2011), Iran (Najmeh and Hamid 2014), Sri Lanka (Bambaradeniya 2016)
135. *Duvaliomimus* – New Zealand (Larochelle and Larivière 2013)
136. *Dyscarius* – Madagascar, Africa, France (Hogan 2012)
137. *Dyscherus* – Switzerland (Erwin and Halpern 1978)
138. *Dyschirius* – United Kingdom (Parry 1975, Speight 1975, Lindroth

1974), Switzerland (Erwin and Halpern 1978), Denmark (Lindroth 1985), Mauritania, Senegal (Basilewsky 1956, 1961a, 1964, 1969), India (Saha et al. 1992), Mexico, South America (Bousquet and Larochelle 1993), Bulgaria, Russia (Guéorguiev and Guéorguiev 1995, Kryzhanovskij et al. 1995); Iran, Caucasia, Transcaucasia, Middle Asia, Siberia (Ghahari et al. 2009), Madagascar (Hogan 2012)

139. *Ecoptomenus* – West and Central African species, Senegal (Basilewsky 1968), India (Saha et al. 1992)
140. *Egadroma* – Distributed through the sub-Saharan northern half of Africa, Senegal (Basilewsky 1951, 1969) and Mauritania (Basilewsky 1964), South Asia, Australia, Pacific (Darlington 1970), New Zealand (Larochelle and Larivière 2013)
141. *Egaploa* – Distributed through almost the whole of sub-Saharan Africa, a second subspecies lives in tropical Asia (Baehr 2003), Senegal, Gambia, Africa (Basilewsky 1951)
142. *Elaphropus* – India (Manjunathan et al. 1978a, Saraswati 1990, Park et al. 2006), Sri Lanka (Bambaradeniya 2016), Northern part of West Africa (Basilewsky 1948, 1956), Guinea (Bruneaude Mire 1964), Iran, Turkey, Caucesia, Siberia, North Mongolia, Far East, Sakhalin, Korean, Peninsula, Japan, Alaska, Canada (Ghahari et al. 2009), Vietnam (Park et al. 2006)
143. *Elaphrus* – Switzerland (Erwin and Halpern 1978), Denmark (Lindroth 1985), Russia (Kryzhanovskij et al. 1995), Bulgaria (Guéorguiev and Guéorguiev 1995), South Africa (Maddison and Ober 2000), Spain (Gutierrez 2004), United Kingdom, United States of America (Bauer 1974, Anderson 1997, Hogan 2012)
144. *Emphanes* – Central Asia, Central and South Europe, Caucasia and Russia (Guéorguiev and Guéorguiev1995, Kryzhanovskij et al. 1995), Turkey (Casale and Taglianti 1999)
145. *Enceladus* – Venezuela (Hogan 2012), South America (Moore 1972)
146. *Endynomena* – Oriental Region, Polynesia (Darlington 1970)
147. *Epilectus* – Australia (Hogan 2012)
148. *Epomis* – Bulgaria (Guéorguiev and Guéorguiev 1995)
149. *Eucamaragnathus* – Peru (Hogan 2012)
150. *Eucartuser* – Bulgaria (Guéorguiev and Guéorguiev 1995)
151. *Eucolliuris* – USA (Blatchley 1910), Germany (Leng 1920), Algeria, Angola, Egypt, Ethiopia, Morocco, Spain (Balearic Islands), Sudan, West Africa (Benin, Senegal) (http://carabidae.org)
152. *Euproctinus* – Chile, Argentina, South America (http//Creative-common.org)
153. *Euryarthron* –West Africa (Wiesner 2001)
154. *Euryscaphus* – Australia (Hogan 2012)
155. *Euschizomerus* – Sri Lanka (Bambaradeniya 2016), India (Saha et al. 1992), Cambodia, Indonesia (Borneo, Sumatra), Malaysia, Vietnam (http://carabidae.org)

156. *Eustra* – Sri Lanka (Bambaradeniya 2016), China (Guéorguiev, 2014), Switzerland (Erwin and Halpern 1978)
157. *Euthenarus* – Australia, New Zealand (Larochelle and Larivière 2013)
158. *Forcipato*r – Guyana, Brazil (Hogan 2012)
159. *Feronia* – Mexico, South America (Bousquet and Larochelle 1993)
160. *Galerita* – Distributed throughout the whole northern half of sub-Saharan Africa including Senegal (Basilewsky 1953b, 1969), Mexico, South America (Bousquet and Larochelle 1993), USA (Larochelle and Larivière 2003)
161. *Galeritula* – USA (Esau 1968), New York, Maryland, Pannsylvania, Indiana, Iowa, Missouri (Casey 1920), Canada, Kansas, Florida (Leng 1920), Vietnam (Park et al. 2006)
162. *Geodromus* – Senegal (Basilewsky 1951)
163. *Geoscaptus* – Australia (Hogan 2012)
164. *Gehringia* – Switzerland (Erwin and Halpern 1978)
165. *Glenopterus* – Frnace (Heer 1847)
166. *Glyptogrus* – Paraguay, Asuncion, Brazil (Hogan 2012), Switzerland (Erwin and Halpern 1978)
167. *Gnaphon* – India, Shambagunur (Hogan 2012), Sri Lanka (Andrews 1929)
168. *Gnathaphanus* – India (Saha et al. 1992, Kumar and Rajagopal 1997), Australia, New Guinea (Darlington 1970), New Zealand (Larochelle and Larivière 2013), Sri Lanka (Bambaradeniya 2016)
169. *Gnathoxys* – Australia (Hogan 2012)
170. *Goniotropis* – Switzerland (Erwin and Halpern 1978)
171. *Gourlayia* – New Zealand (Larochelle and Larivière 2013)
172. *Graniger* – Bulgaria (Guéorguiev and Guéorguiev 1995)
173. *Graphipterus* – West African sub-species recorded from Senegal, Mauritania, Mali, Niger (Basilewsky 1977)
174. *Hakaharpalus* – New Zealand (Larochelle and Larivière 2013)
175. *Haplanister* – New Zealand (Larochelle and Larivière 2013)
176. *Haplogaster* – Switzerland (Erwin and Halpern 1978), India, Bengal (Hogan 2012)
177. *Haplotrachelus* – South Africa, Bedford, Durban (Hogan 2012)
178. *Haptoderus* – Spain (Gutierrez 2004)
179. *Harpaglossus* – Sri Lanka (Bambaradeniya 2016), India (Saha et al. 1992)
180. *Harpalomimete* – Vietnam (Park et al. 2006)
181. *Harpalus* – USA (Everly 1938, Slough 1940, Cornic 1973, Sunderland et al. 1995b), Hungary (Horvatovich and Szarukan 1981), Mexico, South America (Bousquet and Larochelle 1993), Bulgaria (Guéorguiev and Guéorguiev 1995), Russia (Kryzhanovskij et al. 1995), New Zealand (Jorgensen and Lovei 1999), United Kingdom (Fowler 1887, Speight et al. 1982, Atty's 1983, Hyman and Parsons 1992, Luff 1996a, Anderson et

al. 1997), India (Kumar and Prasad 1997), Denmark (Lovei et al. 2000), Spain (Gutierrez 2004), Balkan, Peninsula, Russia, Turkey, Caucasia, Kajakhstan, Uzbekistan, West Africa, Europe, Eukraine, Crimea, Turkey, Caucasia, Iran, Maldova, Ajerbaijan, Daghestan, Armenia, Tarkmenistan, Tajikistan, South Siberia, China, North Korea (Ghahari et al. 2009), Europe, Asia Minor, Mt. Caucasus, Iran, Turkmenistan, Algeria, Central and Balkan Peninsula, Serbia (Carcic and Stankovic 2011), New Zealand (Larochelle and Larivière 2013), Iran (Najmeh and Hamid 2014), Laos, China (Kataev 2015), Austria (Anjum-Zubair 2015)

182. *Helluomorphoides* – New Maxico, South America (Bousquet 2012), North America (Ball 1956)
183. *Hemiaulax* – Cambodia (Kataev 2015)
184. *Heteropaussus* – Sri Lanka (Bambaradeniya 2016)
185. *Hiletus* – Ivory Coast, Mozambique (Hogan 2012)
186. *Holcaspis* – New Zealand (Larochelle and Larivière 2013)
187. *Holcoderus* – Sri Lanka (Bambaradeniya 2016), S.E. Asia to Philippines, New Guinea, Australia, Palau, Caroline (Darlington 1970)
188. *Hololeius* – Sri Lanka, India (Saha et al. 1992), Vietnam (Park et al. 2006), Palaearctic Region (Japan, East China), Oriental Region (Ceylon, India, South China, Taiwan, Philippines, Malaysia, Indonesia), Australian Region (New Guinea, northeast and southeast Australia) (http://eol.org)
189. *Homaloderodes* – Chile (Maddison and Ober 2000)
190. *Hygranillus* – New Zealand (Larochelle and Larivière 2013)
191. *Hyparpalus* – Distributed throughout the whole of sub-Saharan Africa (Basilewsky 1951), Senegal (Lecordier 1988, Basilewsky 1965)
192. *Hyphaereon* – Sri Lanka (Bambaradeniya 2016)
193. *Hypharpax* – New Zealand (Larochelle and Larivière 2013)
194. *Hypherpes* – Mexico, South America (Bousquet and Larochelle 1993)
195. *Ingevaka* – New Zealand (Larochelle and Larivière 2013)
196. *Iniopachys* – Spain (Gutierrez 2004)
197. *Jansenia* – Sri Lanka (Bambaradeniya 2016)
198. *Kaveinga* – New Zealand (Larochelle and Larivière 2013)
199. *Kenodactylus* – New Zealand (Larochelle and Larivière 2013)
200. *Kettlotrechus* – New Zealand (Larochelle and Larivière 2013)
201. *Kiwiharpalus* – New Zealand (Larochelle and Larivière 2013)
202. *Kiwitachys* – New Zealand (Larochelle and Larivière 2013)
203. *Kiwitrechus* – New Zealand (Larochelle and Larivière 2013)
204. *Kupeharpalus* – New Zealand (Larochelle and Larivière 2013)
205. *Kupetrechus* – New Zealand (Larochelle and Larivière 2013)
206. *Kupeus* – New Zealand (Larochelle and Larivière 2013)
207. *Laccopterum* – Australia, Moreton Bay (Hogan 2012)
208. *Lachnothorax* – Sub-Saharan Africa (Basilewsky 1962a), Senegal (Basilewsky 1960b, Liebke 1931) Sri Lanka (Bambaradeniya 2016), India (Saha et al. 1992)

209. *Laemostenus* – Iran, Mediterranean country, Palearctic Region (Ghahari et al. 2009), India (Saha et al. 1992), all continents on Earth, except Antarctica (https://en.wikipedia. org/wiki/ Laemostenus), Bulgaria (Guéorguiev and Guéorguiev 1995)
210. *Lamprias* – South west Asia, Europe, Caucasia, Maldova, Russia, Siberia (Gueorguiev and Gueorguiev 1995, Kryzhanovskij et al. 1995, Neculiseanu and Matalin 2000), Spain (Gutierrez 2004), Turkey (Sahlberg 1912-1913, Kocatepe and Mergen 2004, Kesdek and Yildirim 2007b)
211. *Lasiocera* – Northern part of sub-Saharan Africa, Senegal (Basilewsky 1953a, 1964, 1969), India (Saha et al. 1992), Sri Lanka (Bambaradeniya 2016)
212. *Lebia* – USA, Canada (Madge 1967, Esau 1968), Vietnam, Nepal (Krischenhofer 2010) United Kingdom (Fowler 1887, Britten 1943, Lindroth 1954, Crowson and Crowson 1963), India (Saha et al. 1992), Mexico, South America (Bousquet and Larochelle 1993), Africa, (Kirschenhofer 2010), Sri Lanka (Bambaradeniya 2016), Russia (Kryzhanovskij et al. 1995), Australia (http://carabidae.org/taxa/lebia)
213. *Lecanomerus* – New Zealand (Larochelle and Larivière 2013)
214. *Leistus* – Australia (Lindroth 1974), Switzerland (Erwin and Halpern 1978), United Kingdom (Speight et al. 1982, Jobe 1990), Armenia, Russia, (Kryzhanovskij et al. 1995), Bulgaria (Guéorguiev and Guéorguiev 1995), Denmark (Lindroth, 1985), Spain (Gutierrez 2004), Turkey (Casale and Taglianti 1999), Vietnam (Park et al. 2006), Russia, Caucasia (Ghahari et al. 2009), Iran (Najmeh and Hamid 2014)
215. *Lesticus* – Vietnam (Park et al. 2006), New Guinea and Australia (Darlington 1971)
216. *Leleuporella* – Sri Lanka (Bambaradeniya 2016), India (Abhitha and Sabu 2009)
217. *Licinus* – United Kingdom (Hyman and Parsons 1992), Russia (Kryzhanovskij et al. 1995), Spain (Gutierrez 2004), Iran, Paearctic Region (Ghahari et al. 2009)
218. *Limodromus* – Eurasia (Curcic and Stankovic 2011)
219. *Lionychus* – United Kingdom (Fowles 1989, Fowles and Boyce 1992), Bulgaria (Guéorguiev and Guéorguiev 1995), Sri Lanka (Bambaradeniya 2016)
220. *Lissopterus* – Falkland Islands, South America (Hogan 2012)
221. *Lophoglossus* – Mexico,South America (Bousquet and Larochelle 1993)
222. *Lophyra* – Sri Lanka (Bambaradeniya 2016),West Africa (Baehr 2003), Iran, Palearctic Region (Ghahari et al. 2009)
223. *Loricera* – Switzerland (Erwin and Halpern 1978), Denmark (Lindroth 1985), Belgium (Loreau 1984), United Kingdom (Bauer and Kredler 1988), USA (Symondson and Williams 1995), Bulgaria (Guéorguiev and Guéorguiev 1995), Russia (Kryzhanovskij et al. 1995), Spain (Gutierrez 2004)

224. *Loxandrini* – New Zealand (Larochelle and Larivière 2013)
225. *Loxandrus* – Mexico, South America (Bousquet and Larochelle 1993)
226. *Loxocrepis* – Sri Lanka (Bambaradeniya 2016)
227. *Loxomerus* – New Zealand (Larochelle and Larivière 2013)
228. *Loxoncus* – West Africa, Senegal, Gambia (Basilewsky 1951), Sri Lanka (Bambaradeniya 2016)
229. *Luperca* – India (Hogan 2012), Ceylon and Burma (Andrewes 1929)
230. *Lymnastis* – West and Central Africa (Basilewsky 1961a), Mauritania (Basilewsky 1964, 1970), and Senegal (Basilewsky1969), India (Saha et al. 1992), United States of America, (Bousquet, 2012), Malayasia (Lim and Pan 1980), Vietnam (Park et al. 2006)
231. *Macrocheilus* – Sri Lanka (Perera 1981), India (Kumar and Rajagopal 1997, Raj et al. 2012), Hongkong, China, Macau, Sumatra, Guinea Bissau, Senegal (http://carabidae.org/taxa/macrocheilus-hope)
232. *Macromorphus* – United Kingdom (Hogan 2012), USA (Erwin and Ball 1982)
233. *Madascaris* – United States of America (Hogan 2012), Africa (Basilewsky 1980)
234. *Mamboicus* – Zambia, Lusaka,Tanzania Morogoro (Hogan 2012), Switzerland (Erwin and Halpern 1978)
235. *Maoriharpalus* – New Zealand (Larochelle and Larivière 2013)
236. *Maoripamborus* – New Zealand (Larochelle and Larivière 2013)
237. *Maoritrechus* – New Zealand (Larochelle and Larivière 2013)
238. *Masoreus* – United Kingdom (Lindroth 1974), Sri Lanka (Bambaradeniya 2016), Mexico, South America (Bousquet and Larochelle 1993), Russia (Kryzhanovskij et al.1995)
239. *Mastax* – United Kingdom (Lindroth 1974), Bulgaria, (Guéorguiev and Guéorguiev 1995),Vietnam (Park et al. 2006)
240. *Mecodema* – New Zealand (Larochelle and Larivière 2013)
241. *Mecyclothoracini* – New Zealand (Larochelle and Larivière 2013)
242. *Mecyclothorax* – New Zealand (Larochelle and Larivière 2013)
243. *Mecynoscaris* – Madagascar, Mt. De Ambre (Hogan 2012), Switzerland (Erwin and Halpern 1978)
244. *Megacephala* – South America (Choate 2003), Australia (Kromp 1990)
245. *Megadromus* – New Zealand (Larochelle and Larivière 2013)
246. *Melaenus* – Sri Lanka (Bambaradeniya 2016), India (Kumar and Rajagopal 1997, Kushwaha and Hegde 2012)
247. *Melanchiton* – Northern part of Central and East Africa (Straneo 1950)
248. *Melanospilus* – Sri Lanka (Bambaradeniya 2016)
249. *Melisiodera* – Australia, Victoria (Hogan 2012)
250. *Menigius* – Cameroon, Sardi (Hogan 2012), Switzerland (Erwin and Halpern 1978)
251. *Meonis* – Australia, Sydney (Hogan 2012)
252. *Meotachys* – South America (Erwin 1974)
253. *Merizodus* – Chile, South America (Maddison and Ober 2000)

254. *Merizomena* – United Kingdom (Lindroth 1974), Russia, Kazakhstan, Turkmenistan, Uzbekistan, Tajikistan (Ghahari et al. 2009)
255. *Mesus* – Switzerland (Erwin and Halpern 1978)
256. *Metabletus* – United Kingdom (Fowler 1887, Speight 1982), India (Saha et al. 1992)
257. *Metazuphium* – Sri Lanka (Bambaradeniya 2016)
258. *Metrius* – Switzerland (Erwin and Halpern 1978)
259. *Micratopus* – USA, Arizona, South America (Maddisona and Ober 2000)
260. *Microlestes* – United Kingdom (Lindroth 1974), Africa, India (Saha et al. 1992), Bulgaria (Guéorguiev and Guéorguiev 1995), Spain (Gutierrz 2004), Vietnam (Park et al. 2006)
261. *Mimocolliuris* – Sri Lanka (Bambaradeniya 2016), Germany (Lorenz 2005)
262. *Mioptachys* – Mississippi, USA (Maddison and Ober 2000)
263. *Miscelus* – Sri Lanka (Bambaradeniya 2016), China, Indonesia (Java, Sumatra), India (Sikkim, Darjeeling), Melanesia (Papua), Philippines Formosa), Taiwan (http://carabidae.org/taxa/javanus-klug)
264. *Miscodera* – Denmark (Lindroth 1985), Europe, United Kingdom (Alexander 1993)
265. *Mochtherus* – Sri Lanka (Bambaradeniya 2016), Indonesia (Krischenhofer 2010)
266. *Molops* – Balkan Peninsula, Europe (Carcic and Stancovic 2011)
267. *Molopsida* – New Zealand (Larochelle and Larivière 2013)
268. *Monocentrum* – Australia, Cairns (Hogan 2012), USA (Erwinand Ball 1982)
269. *Moriomorphina* – New Zealand (Larochelle and Larivière 2013)
270. *Morion* – Sri Lanka (Bambaradeniya 2016), Iran, India, Turkey (Casale and Taglianti 1999), Caucasia, Armenia, Russia, Iran (Kryzhanovskij et al. 1995, Ghahari et al. 2009), South Asia, Philippines, Moluccas (Darlington 1970), Mexico, South America (Hill 1993)
271. *Mouhotia* – Thailand (Hogan 2012), Switzerland (Erwin and Halpern 1978)
272. *Myas* – Bulgaria (Guéorguiev and Guéorguiev 1995)
273. *Myriochila* – Sri Lanka (Bambaradeniya 2016), Iran, Palearctic Region (Ghahari et al. 2009)
274. *Mystropomus* – Switzerland (Erwin and Halpern 1978)
275. *Nanodiodes* – Sri Lanka (Bambaradeniya 2016), Australia, New Zealand (Will 2015)
276. *Neanops* – New Zealand (Larochelle and Larivière 2013)
277. *Nebria* – North America (Erwin 1985), Denmark (Lindroth 1985), Europe (Sopp and Wratten 1986), United Kingdom (Gilbert 1946, Nelemans 1987a and b, King and Stabins 1971, Constantine 1993),

Bulgaria (Guéorguiev and Guéorguiev 1995), USA (Burgess et al. 2002), Spain (Gutierrez 2004), Asia Minor; Caucasia, Bulgaria, Iran, Israel, Syria, Armenium, Russia (Kryzhanovskij et al. 1995, Löbl and Smetana 2003, Ghahari et al. 2009), Turkey (Csiki 1927b; Öncüer 1991, Casale and Taglianti 1999, Avgin and Emre 2007, Anlas and Tezcan 2010, Tezcan et al. 2011), Europe, Asia Minor, Mt. Caucasus, Iran, North America (introduced), Serbia (Curcic and Stankovic 2011), Austria (Anjum-Zubair 2015)

278. *Neocarenum* – Australia, Perth (Hogan 2012)
279. *Neochryopus* – Nigeria, Oban (Hogan 2012), West and Central Africa (Basilewsky 1968)
280. *Neocollyris* – Sri Lanka (Bambaradeniya 2016), India (Saha and Halder 1986), Philippines (Santos 2014)
281. *Neoferonia* – New Zealand (Larochelle and Larivière 2013)
282. *Nesamblyops* – New Zealand (Larochelle and Larivière 2013)
283. *Nomius* – USA, Arizona (Maddison and Ober 2000),
284. *Notagonum* – New Zealand (Larochelle and Larivière 2013)
285. *Notaphus* – Denmark (Lindroth 1985), Australia, Bulgaria, Siberia, Caucasia, Czech Republic, Russia (Guéorguiev and Guéorguiev 1995, Kryzhanovskij et al. 1995, Hurka 1996), Turkey (Sahlberg 1912-1913, Kesdek and Yildirm 2007a), Mexico, South America (Bousquet and Larochelle 1993), USA (Bousquet 2012)
286. *Notiobia* – New Zealand (Larochelle and Larivière 2013)
287. *Notiophilus* – Belgium (Loreau 1984), United Kingdom (Anderson 1972, Luff 1981b Alexander 1993), Mexico, South America (Bousquet and Larochelle 1993), Bulgaria (Guéorguiev and Guéorguiev 1995), United States of America (Larochelle and Larivière 2003), Spain (Gutierrez 2004)
288. *Oarotrechus* – New Zealand (Larochelle and Larivière 2013)
289. *Ochryopus* – Uganda, Makerere (Hogan 2012), Czech Republic, Slovakia, Russia (Fedorenko, 2014)
290. *Ocydromus* – Spain (Gutierrez 2004)
291. *Odacantha* – India (Saha 1992), Russia (Kryzhanovskij et al. 1995), Europe, Caucasia, Siberia, Russia (Ghahari et al. 2009)
292. *Oedesis* – Bulgaria (Guéorguiev and Guéorguiev 1995), Uzbekistan (Kataev 2015)
293. *Olisthopus* – United Kingdom (Lindroth 1974), Mexico, South America (Bousquet and Larochelle 1993), Europe to Central Italy and East to Asia Minor and the Caucasus (http://www.habitas.org.uk.)
294. *Omalodera* – Chile (Maddison and Ober 2000)
295. *Omphra* – Sri Lanka (Bambaradeniya 2016), India, Nepal, Sri Lanka (Raj et al. 2012)
296. *Omphron* – West Africa, Senegal (CSIKI 1927, Basilewsky1969),United Kingdom, South India (Prasad and Rajagopal 1990, Rajagopal and

Kumar 1992, Lognathan and David 1999, Raj et al. 2012), Vietnam (Park et al. 2006), Denmark (Lindroth 1985)

297. *Onawea* – New Zealand (Larochelle and Larivière 2013)
298. *Oodes* – United Kingdom (Lindroth 1974), India (Saha 1992), Bulgaria (Guéorguiev and Guéorguiev 1995), Russia (Kryzhanovskij et al. 1995), Central and South Europe, Turkey, Caucasia, Turmenistan, Iran (Ghahari et al. 2009)
299. *Ooidius* – Sri Lanka (Bambaradeniya, 2016), Africa (Basilewsky 1950-51)
300. *Oopterus* – New Zealand (Larochelle and Larivière 2013)
301. *Oosoma* – Sri Lanka (Bambaradeniya 2016)
302. *Ophionea* – India (Samal and Misra 1984), Sri Lanka (Bambaradeniya 2016), Vietnam (Park et al. 2006)
303. *Ophoniscus* – Sri Lanka (Bambaradeniya 2016), Vietnam, Afghanisthan (Kataev 2015)
304. *Ophonus* – United Kingdom (Lindorth 1974), Bulgaria (Guéorguiev and Guéorguiev 1995), Spain (Gutierrez 2004), Mediterrranean Countries, Ukraine, Russia, Maldova, Crimea, Ajerbaijan, Daghestan, Armenia, Kazakhstan (Ghahari et al. 2009), Europe, Asia Minor, Iran, Iraq, Israel, Syria, Serbia (Curcic and Stakovic 2011), Iran (Najmeh and Hamid 2014), Austria (Anjum -Zubair 2015)
305. *Oregus* – New Zealand (Larochelle and Larivière 2013)
306. *Orthoglymma* – Australia, New Zealand (Larochelle and Larivière 2013)
307. *Orthogonius* – West and Central Africa, Senegal, (Basilewsky 1968), Sri Lanka (Bambaradeniya 2016), Vietnam (Park et al. 2006)
308. *Orthotrichus* – Iran, Plearctic Region, Europe, Mountains of SE Middle Asia, Turkey, Russia, Moldova, Ukraine, Armenia (Ghahari et al. 2009)
309. *Oxycentrus* – India (Saha et al. 1992) Taiwan, Japan (http://ecoregister.org/)
310. *Oxylobus* – India, Nilgiri hills (Hogan 2012, Rajagopal and Kumar 1992, Kumar, 1997), Sri Lanka (Bambaradeniya 2016)
311. *Ozaena* – Switzerland (Erwin and Halpern 1978)
312. *Pachydesus* – South Africa (Maddison and Ober 2000)
313. *Pamborus* – Australia (Takami and Sota 2006), Switzerland (Erwin and Halpern 1978)
314. *Panagaeus* – North Africa, Caucasia, Siberia, Bulgaria, Iran, Maldova, Russia (Guéorguiev and Guéorguiev 1995, Kryzhanovskij et al. 1995, Neculiseanu and Mathalin 2000) Turkey (Yücel and Sahin 1988), United Kingdom (Speight et al. 1982), Turkey, Iran Caucesia (Ghahari et al. 2009)
315. *Parabaris* – New Zealand (Larochelle and Larivière 2005, 2013)
316. *Parabradycellus* – Vietnam (HauGiang), China, Thailand (Park et al. 2006)
317. *Paranchus* – Spain (Gutierrez 2004), Austria, Azores, Belgium, Bosnia and Herzegovina, Bulgaria, Byelorussia Canary

Isles Croatia, Denmark, Estonia, Finland, France, Great Britain, Germany, Hungary, Ireland, Italy, Latvia, Lithuania, Malta, Morocco, Madeira Arch., Netherlands, Norway, Poland, Russia, Slovakia, Slovenia, Spain, Sweden, Switzerland, Turkey, Tunisia, Ukraine (http://carabidae.org/taxa/albipes-fabricius)

318. *Paraphaea* – Andaman isl., Myanmar (Burma), Cambodia, China, Indonesia (Borneo Java, Sumatra), Japan, Melanesia (PNG), Micronesia (MRN), Philippines, Thailand, Taiwan (Formosa), Vietnam (http://carabidae.org/taxa/ binotata-dejean), Sri Lanka (Bambaradeniya 2016)
319. *Paraphonus* – Bulgaria (Guéorguiev and Guéorguiev 1995), Spain (Gutiérrez 2004), Sri Lanka (Bambaradeniya 2016)
320. *Paratachys* – New Zealand (Larochelle and Larivière 2005)
321. *Parazuphium* – Iran (Najmeh and Hamid 2014), Germany (Lorenz 2005)
322. *Parena* – India (Rao et al. 1971, Rao 1978, Pillai and Nair 1990, Puspalatha and Veeresh 1995, Kumar and Rajagopal 1997), Bulgaria (Guéorguiev and Guéorguiev 1995), Sri Lanka (Bambaradeniya 2016), South Africa and South East Asia (http.//creative-commons.org)
323. *Pasimachus* – United States of America, Mexico, Jalapa (Hogan 2012), Switzerland (Erwin and Halpern 1978)
324. *Passalidius* – South Africa (Hogan 2012), USA (Ball and Erwin 1982)
325. *Patrobus* – United Kingdom (Speight et al. 1982, Houston and Luff 1983, Allen 1991), India (Saha et al. 1992), Russia (Kryzhanovskij et al. 1995)
326. *Paussus* – Senegal (Basilewsky 1965, 1969), Russia (Kryzhanovskij et al. 1995), Sri Lanka (Bambaradeniya 2016)
327. *Peliocypa*s – Sri Lanka (Bambaradeniya 2016), Taiwan (http://taibif.tw/en/catalogue)
328. *Pelodiaetodes* – New Zealand (Larochelle and Larivière 2013)
329. *Pelocharis* – Sri Lanka (Bambaradeniya 2016)
330. *Pelophila* – United Kingdom (Lindroth 1974), Switzerland (Erwin and Halpern 1978), Denmark (Lindroth 1985)
331. *Pentagonica* – Sri Lanka (Bambaradeniya 2016), South Asia, Australia (Darlington 1970), Russia (Kryzhanovskij et al. 1905), New Zealand (Larochelle and Larivière 2013)
332. *Pericompsus* – United States of America, Arizona (Maddison and Ober 2000), New Zealand (Larochelle and Larivière 2013)
333. *Perigona* – United Kingdom (Hinton 1945, Allen 1950b), South Asia, New Guinea (Darlington 1970), India (Saha et al. 1992), Bulgaria (Guéorguiev and Guéorguiev 1995), Russia (Kryzhanovskij et al. 1995)
334. *Perileptus* – United Kingdom (Lindroth 1974), Denmark (Lindroth 1985), Armenia, Azerbaijan, Georgia, Iran, Iraq, Israel, Saudi Arabia, Syria, Turkey and in African nations of Tunisia and Canary

Islands (https://en.wikipedia.org/wiki/Perileptus), Spain, Madrid (Maddison and Ober 2000)

335. *Peripristus* – Sri Lanka (Bambaradeniya 2016), India (Saha et al. 1992)
336. *Pheropsophus* – West African and northern sub-Sahara, Senegal (Basilewsky 1969) and Mauritania (Basilewsky 1970), North Eastern India (Atwal 1986), Philippines (Sigsgaard et al. 1999) China, Mayanmar, Thailand, Vietnam (Krischenhofer 2010), Russia, Bulgaria, Iran (Ghahari et al. 2009), West Africa (Woin et al. 2005), South America (Frank et al. 2009), Vietnam (Park et al. 2006)
337. *Philorhizus* – Brtish Isles, France, Portugal and Spain, distributed from west and central Europe south to north Africa, Asia Minor and the Caucasus (http://www.habitas. org. uk/groundbeetles), Spain (Gutierrz 2004)
338. *Philoscaphus* – Australia, New South Wales (Hogan 2012)
339. *Pholeodytes* – New Zealand (Larochelle and Larivière 2013)
340. *Phloeozeteus* – Switzeland (Erwin and Halpern 1978)
341. *Phosphodrus* – New Zealand (Larochelle and Larivière 2013)
342. *Phrypeus* – USA, California (Maddison and Ober 2000)
343. *Physolaesthus* – Australia, New Zealand (Larochelle, and Larivière, 2013)
344. *Planetes* – Sri Lanka (Bambaradeniya 2016), India (Saha et al. 1992)
345. *Platyderus* – United Kingdom (Speight 1982), Spain (Gutierrez 2004), Sri Lanka (Bambaradeniya 2016)
346. *Platymetopus* – West and Central Africa (Basilewsky 1950a), Senegal (Basilewsky-1950a, 1961a, 1965), India (Saha et al. 1992), Vietnam (Park et al. 2006)
347. *Platynus* – Russia (Kryzhanovskij et al. 1995), Spain (Gutierrez 2004), USA (Bousquet and Larochelle 1993)
348. *Platysma* – Mexico, South America (Bousquet and Larochelle 1993)
349. *Plocamostethus* – New Zealand (Larochelle and Larivière 2013)
350. *Poecilus* – Mexico, South America (Bousquet and Larochel 1993), Bulgaria (Guéorguiev and Guéorguiev 1995), United States of America (Larochelle and Larivière 2003), Europe, England (Heyler et al. 2009), Austria (Anjum-Zubair 2015), Germany (Lang 2003), Spain (Gutierrez 2004), Central Asia, Europe, Turkey, Iran, Caucasia, Syria, Siberia (Ghahari et al. 2009), Europe, Siberia, Northeast China, Asia Minor, Mt. Caucasus, Syria, Central Asia, Serbia (Curcic and Stankovic 2011), Spain (Gutierrez 2004)
351. *Pogonistes* – Bulgaria (Guéorguiev and Guéorguiev 1995)
352. *Pogonus* – Senegal (Basilewsky 1964), Denmark (Lindroth,1985), whole of Africa and Madagaskar, likewise in the Mediterranean,Chad (Bruneau De Mire 1990), United Kingdom (Hyman and Parsons 1992),Russia (Kryzhanovskij et al. 1995), Bulgaria (Guéorguiev and Guéorguiev 1995)

353. *Polistichus* – United Kingdom (Henderson 1991), Bulgaria (Guéorguiev and Guéorguiev 1995), Iran, Russia, Caucasia (Ghahari et al. 2009)
354. *Polyderis* – Whole of Africa (Basilewsky 1948), Mauritania (Basilewsky 1964), Senegal (Basilewsky 1953b, 1956), Vietnam (Park et al. 2006), USA, Arizona (Maddison and Ober 2000), New Zealand (Larochelle and Larivière 2013)
355. *Polystichus* – Bulgaria (Guéorguiev and Guéorguiev 1995)
356. *Porotachys* – Azerbajan, Algeria, Armenia, Bulgaria, Byelorussia, Czech, Estonia, Finland, Great Britain, Georgia, Greece, Hungary, Italy, Latvia, Lithuania, Moldavia, Zerbajan, Morocco, Netherlands, Poland, Romania, Russia, Slovakia, Spain, Sweden, Switzerland, Ukraine (http://carabidae.org/taxa/bisulcatus-nicolai), Bulgaria (Guéorguiev and Guéorguiev 1995)
357. *Progonochaetus* – Distributed throughout the sub-Saharan Africa (Basilewsky 1950a, 1956), Senegal, Gambia (Basilewsky 1956, 1969, 1950a), Pakistan (Kataev 2012)
358. *Promecognathus* – United States of America, California (Hogan 2012)
359. *Prosphodrus* – New Zealand (Larochelle and Larivière 2013)
360. *Prosopogmus* – Australia, New Zealand (Larochelle and Larivière 2013)
361. *Prothyma* – Senegal (Basilewsky 1969), West Africa (Wiesner 2001)
362. *Protocollyris* – Sri Lanka (Bambaradeniya 2016),
363. *Psegmatopterus* – New Zealand (Larochelle and Larivière 2013)
364. *Pseudamara* – Mexico, South America (Bousquet and Larochelle 1993)
365. *Pseudocalleida* – Malaysia (Krischenhofer 2010)
366. *Pseudoclivina* – Mauritania (Basilewsky 1970), Senegal (Basilewsky 1965)
367. *Pseudognathaphanus* – Sri Lanka (Bambaradeniya 2016), Vietnam (Park et al. 2006)
368. *Pseudophonus* – Bulgaria (Guéorguiev and Guéorguiev 1995), Spain (Gutierrez 2004)
369. *Pseudorhysopus* – Guangxi, China, Laos (Kataev 2015)
370. *Pseudozaena* – Philippines, New Guinea (Darlington 1970)
371. *Psydrus* – Russia (Kryzhanovskij et al. 1995), USA, New Mexico (Maddison and Ober 2000), United States of America, California (Hogan 2012)
372. *Pterostichus* – United Kingdom (Moore 1957b, Boer den 1977, Speight et al. 1982, Luff 1989, Hyman and Parsons 1992, Duff 1993, Anderson 1993), Mexico, South America (Bousquet and Larochelle 1993), Bulgaria (Guéorguiev and Guéorguiev 1995), Russia (Kryzhanovskij et al. 1995), Mexico, South America (Bousquet and Larochelle 1993, Larochelle and Larivière 2003), Spain (Gutierrez 2004), Central Asia, Europe, Turkey, Caucasia, Iran, Siberia (Ghahari et al. 2009), Europe, Mt. Caucasus, Siberia, Transbaikal area, Japan, East Palaearctic, Balkan

Peninsula, Asia Minor, Serbia (Curcic and Stankovic 2011), Austria (Anjum-Zubair 2015)

373. *Risophilus* – East Angila, United Kingdom (Davis 1973), Vietnam (Park et al. 2006)
374. *Salcedia* – Switzerland (Erwin and Halpern 1978)
375. *Scaphinotus* – Mexico, South America (Bousquet and Larochelle 1993)
376. *Scapterus* – North West India (Hogan 2012)
377. *Scaraphites* – Australia, Sydney (Hogan 2012)
378. *Scarites* – West Africa (Benninger 1938), Colombia, Brazil, France, United States of America, Florida, Saudi Arabia, Algeria, Japan, Jambia, Lusaka (Hogan 2012), Middle of Asia, Caucasia, Russia, Siberia, Armenia, Crimea, Kazakhastan, Maldova, Ukraine, Uzbekistan (Kryzhanovskij et al. 1995), North Carolina, USA (Esau 1968, Laub and Luna 1992), Europe, Algeria, Egypt, Libya, Tunisia, Afghanistan, Cyprus, India, Iraq, Iran, Israel, Kazakhastan, Pakistan, Saudi Arbia, Syria, Turkmenistan, Uzbekistan, Yemen, Syria (Kryzhanovskij et al. 1995, Löbl and Smetana 2003, Ghahari et al. 2009), North Africa (Libya),China, Crimea, Hungary, Mongolia, Tadjikistan, Bulgaria, Ukraine (Guéorguiev and Guéorguiev 1995), Turkey (Sahlberg 1912-1913, CSIKI 1927b, Öncüer 1991, Casale and Taglianti 1999), India (Saha et al. 1997, Kumar 1997), USA (Larochelle and Larivière 2003), Vietnam (Park et al. 2006), Switzerland (Erwin and Halpern 1978)
379. *Schizogenius* – United States of America (Hogan 2012), Switzerland (Erwin and Halpern 1978)
380. *Scopodes* – Australia, New Zealand (Larochelle and Larivière 2013)
381. *Scototrechus* – Australia, New Zealand (Larochelle and Larivière 2013)
382. *Scybalicus* – United Kingdom (Allen 1989)
383. *Siagona* – Switzerland (Erwin and Halpern 1978), Russia (Kryzhanovskij et al. 1995)
384. *Sinechostichus* – Italy,Tuscany (Maddison and Ober 2000)
385. *Siopelus* – West and Central Africa, Mauritania (Basilewsky 1964), Senegal (Basilewsky 1950a, 1964)
386. *Selenophorus* – Central America, Polynesia (Darlington 1970), USA (Blachley 1910, Larochelle and Larivière 2003)
387. *Selina* – Sri Lanka (Bambaradeniya 2016), Malayasia (Lim and Pan 1980)
388. *Sericoda* – Sri Lanka (Bambaradeniya 2016), USA (Dearborn et al. 2014)
389. *Siagona* – Bulgaria (Guéorguiev and Guéorguiev 1995), Sri Lanka, Iran, India, North Africa, East Asia, Europe, Caucasia, Armenia, Crimea, Iraq, Iran, Kazakhstan, Russia, Senegal, Sudan, Tadjikistan, Ukraine, Sri Lanka, Spain, Africa, Mediterranean countries (Ghahari et al. 2009, Hogan 2012, Saha et al. 1992), South America (Moore 1972)
390. *Sitaphe* – Australia (Maddison and Ober 2000)

391. *Sloaneana* – Australia, Tasmania (Maddison and Ober 2000)
392. *Sofota* – Malaysia (Krischenhofer 2010)
393. *Solenogenys* – United Kingdom (Lindroth 1974), USA (Bousquet 2012)
394. *Somoplatus* – West and Central Africa, Senegal (Basilewsky 1969, 1986), India (Saha et al. 1992)
395. *Somotrichus* – Sri Lanka (Bambaradeniya 2016), Cosmopolitian (Darlington 1970)
396. *Sphaerotachys* – Whole of Africa, Southern Europe, Senegal (Basilewsky 1964)
397. *Sphodrus* – Iran, South and West Asia, Europe, Turkey, Caucasia, India, Conary Island (Ghahari et al. 2009), Russia (Kryzhanovskij et al. 1995), Bulgaria (Guéorguiev and Guéorguiev 1995)
398. *Stenocrepis* – Mexico, South America (Bousquet and Larochelle 1993)
399. *Stenolophus* – Mexico, South America (Bousquet and Larochelle 1993), Russia (Kryzhanovskij et al. 1995), China (Kataev 2015)
400. *Steropus* – Spain (Gutierrez 2004)
401. *Stomis* – United Kingdom (Lindroth 1974), Bulgaria (Guéorguiev and Guéorguiev 1995), Spain (Gutierrez 2004), Europe, Asia Minor, Mt. Caucasus, North America, Serbia (Curcic and Stancovic 2011)
402. *Stomonaxellus* – Sri Lanka (Bambaradeniya 2016)
403. *Storthodontus* – Madagascar, Africa (Hogan 2012), Switzerland (Erwin and Halpern 1978)
404. *Styphlomerus* – Throughout Africa, Senegal (Basilewsky 1965, 1969, Liebke 1934), India (Saha et al. 1992), Sri Lanka (Bambaradeniya 2016)
405. *Syletor* – Sri Lanka (Bambaradeniya 2016)
406. *Syllectus* – Australia, New Zealand (Larochelle and Larivière 2013)
407. *Synteratus* – Australia, New Zealand (Larochelle and Larivière 2013)
408. *Syntomus* – Afghanistan, Algeria, Armenia, Bhutan, Canary Isles, Cyprus, Egypt, France, Georgia, Greece, Iran, India (Himachal Pradesh), Iraq, Israel, Italy, Kyrgyzstan, Kazakhstan, Lebanon, Macedonia, Morocco, Madeira Arch., Pakistan, Portugal, Russia, Spain, Syria, Turkmenistan, Turkey, Tunisia, Uzbekistan (http://carabidae.org/taxa/fuscomaculatus-motschulsky), Bulgaria (Guéorguiev and Guéorguiev 1995), Spain (Gutierrz 2004), Sri Lanka (Bambaradeniya 2016)
409. *Synuchus* – United Kingdom (Lindroth 1985-86), Russia (Kryzhanovskij et al. 1995), Spain (Gutierrez 2004), Iran, Caucasia (Ghahari et al. 2009)
410. *Syrdenus* – Canary Islands, West Mediterranean and in the Northern part of West Africa, Mauritania (Basilewsky 1970), Tschad (Bruneau De Mire 1990), Senegal (Basilewsky 1969)
411. *Systolocranius* – India (Saha et al. 1992)
412. *Tachys* – Cap Verde Islands, Senegal, Mauritania (Basilewsky 1962b), Oriental and Australian regions (Basilewsky 1964, 1969), Zaire (Basilewsky 1953b), USA (Esau 1968) Africa, South Asia, Australia, New Guinea, Polynesia, Singapore, Hawaii, Tahiti, Africa, West Indies

(Darlington 1970), Denmark (Lindroth 1985) India (Saraswati 1990), Russia (Kryzhanovskij et al. 1995), United Kingdom (Edmonds 1934, Moore 1956, Welch 1992, Williams 1997), Vietnam (Park et al. 2006), United States of America, Arizona (Maddisona and Ober 2000)

413. *Tachyt*a – Denmark (Lindroth 1985), Azerbaijan, Albania, Andorra, Antilles islands (Cuba), Armenia, Austria, Belgium, Bosnia and Herzegovina, Belize, Bulgaria, Byelorussia, Canada, China, Croatia, Cyprus, Czech, Denmark, Estonia, Finland, France (Corse), Great Britain, Germany, Georgia, Greece, Guatemala, Hungary, Iran, Ireland, Israel, Italy (Sardegna, Sicily), Japan, Kyrgyzstan, Kazakhstan, Latvia, Liechtenstein, Lithuania, Luxenbourg, Macedonia, Moldavia, Montenegro, Mongolia, Morocco, North Korea, Netherlands, Norway, Poland, Portugal, Romania, Russia, Slovakia, Slovenia, Spain, Serbia, Switzerland, Tadzhikistan, Turkmenistan, Turkey, United States, Uzbekistan (http://carabidae.org/taxa/nana-gyllenha)
414. *Tachyura* – West Africa (Bruneau De Mire 1964, 1963), Mauritania (Basilewsky 1964), Senegal (Basilewsky 1950b, 1964, 1968, 1969), Vietnam (Park et al. 2006)
415. *Taeniolobus* – United Kingdom (Lindroth 1974), Switzerland (Erwin and Halpern 1978)
416. *Taenorthrus* – Australia, New Zealand (Larochelle and Larivière 2013)
417. *Taicona* – Vietnam (Park et al. 2006)
418. *Tangarona* – Australia, New Zealand (Larochelle and Larivière 2013)
419. *Taphoxenu*s – Iran, Russia, Kazakhastan, Tajikistan, Turkmenistan, Uzbekistan, Europe (Ghahari et al. 2009), Bulgaria (Guéorguiev and Guéorguiev 1995)
420. *Tarastethus* – Australia, New Zealand (Larochelle and Larivière 2013)
421. *Taridius* – Northeastern India, Malaysia (Krischenhofer 2010)
422. *Tetragonoderus* – West Africa (Basilewsky 1968), Mauritania, Senegal (Basilewsky 1961a, 1964, 1965, 1969, 1970), Russia (Kryzhanovskij et al. 1995), India (Kumar and Rajagopal 1997, Patil and Sathe 2003), United States of America (Larochelle and Larivière 2003)
423. *Thalassophilus* – United Kingdom (Hyman and Parsons 1992), Russia (Maddison and Ober 2000)
424. *Therates* – Philippines, New Guinea (Darlington 1970)
425. *Thlibops* – India, Senegal, India (Hogan 2012, Saha et al. 1992)
426. *Thyreopterus* – Democratic Republic of the Congo (http://carabidae.org/taxa/bilunatus-burgeon)
427. *Tibioscarites* – Switzerland (Erwin and Halpern 1978)
428. *Tonkinoscaris* – Switzerland (Erwin and Halpern 1978)
429. *Trachypachus* – Denmark (Lindroth 1985)
430. *Trechodes* – Australia (Maddison and Ober 2000)
431. *Trechosiella* – South Africa (Maddison and Ober 2000)
432. *Trechus* – Belgium (Loreau 1984), Denmark (Lindroth 1985), United

Kingdom (Boer 1965, Luff 1996a and b, Luff and Wardle 1991, Day 1987, Allen 1950a), Russia (Kryzhanovskij et al. 1995), USA (Maddison and Ober 2000), Spain (Gutierrez 2004), Vietnam (Park et al. 2006)

433. *Trichocellus* – United Kingdom (Luff 1996a), Bulgaria (Guéorguiev and Guéorguiev 1995)
434. *Trichopsida* – Australia, New Zealand (Larochelle and Larivière 2013)
435. *Trichotichnus* – India (Saha et al. 1992), Sri Lanka (Bambaradeniya 2016), Vietnam (Park et al. 2006)
436. *Tricondylla* – Sri Lanka (Bambaradeniya 2016), Japan (Sawada 2001)
437. *Trigonothops* – Australia, New Zealand (Larochelle and Larivière 2013)
438. *Trigonotoma* – Sri Lanka (Bambaradeniya 2016), Vietnam (Park et al. 2006)
439. *Trilophidius* – Sri Lanka (Bambaradeniya 2016), Oriental region (Balkenohl 2001)
440. *Trilophus* – India, Sri Lanka (Bambaradeniya 2016), Oriental region (Balkenohl 1999)
441. *Triplosarus* – Australia, New Zealand (Larochelle and Larivière 2013)
442. *Tropopsis* – Switzerland (Erwin and Halpern 1978)
443. *Tropopterus* – Australia (Bousquet and Larochelle 1993)
444. *Tschitscherinellus* – Afghanistan (Kataev 2015)
445. *Tuiharpalus* – Australia, New Zealand (Larochelle and Larivière 2013)
446. *Typhlocharis* – Spain (Maddison and Ober 2000)
447. *Typhlonesiotes* – Hawaii (Darlington 1970)
448. *Typhloscaris* – Kenya, Tanzania, Tanganyika Range (Hogan 2012), Switzerland (Erwin and Halpern 1978)
449. *Vietotrechus* – Vietnam (Park et al. 2006)
450. *Waiputrechus* – Australia, New Zealand (Larochelle and Larivière 2013)
451. *Xenodochus* – West Africa, Senegal, Gambia (Basilewsky 1950a, 1969)
452. *Zabrus* – United Kingdom (Bassett 1978), Spain (Gutierrez 2004), Caucasi, Russia, Iran, Turkey (Ghahari et al. 2009), Iran (Najmeh and Hamid 2014)
453. *Zeopoecilus* – Australia, New Zealand (Larochelle and Larivière 2013)
454. *Zolotarewskyella* – Africa (Baehr, 2003)
455. *Zolus* – Australia, New Zealand (Larochelle and Larivière 2013)
456. *Zuphium* – Senegal (Basilewsky 1962a, 1969), sub-Saharan Africa, Mauritania (Basilewsky 1970), Philippines (Heinrich et al. 2004), Bulgaria (Guéorguiev and Guéorguiev 1995), Russia (Kryzhanovskij et al. 1995), Iran, India, Europe, Mediterranean countries, England, Russia, Maldova, Ukraine, Armenia, Turkmenistan, Uzbekistan (Ghahari et al. 2009)

Appendix C

Glossary of Terms Used in Applied Entomology and Biological Control

A-AB – Latin prefix, off, from, away, apart

A-An – Greek prefix; not without

Abactinal (Asboral) – The surface of the body opposite to the mouth in an animal showing radial symmetry.

Abdomen –The most posterior of the three main body divisions of insects; it is usually composed of ten or eleven segments and bears no functional legs in the adult stage.

Abdominal feet – The freshly abdominal legs found in caterpillars (Lepidoptera) and sawflies (Hymenoptera) bearing a characteristic arrangement of crochets

Acanth-Acanths – Prefix from Greek *Akanthos*: a thorn, spine

Acari – An alternative name for *Acarina,* ticks and mites

Acephalous – Without a tread. The term is also used to describe creatures whose heads are much reduced or are invaginated into the thorax, as in certain insect larvae.

Acoustical – Communication by sound and hearing

Acr-, Acri-, Acro- – Prefixes from Greek *Akros*: top, point, tip

Acron – Anterior, preoral body cap of a segmented animal

Acrosternite – The portion of the tergum anterior to the antecostal suture

Actin-, Action – Prefix from Greek *Aktis*, *aktinos*: a ray

Active ventilation – Ventilation aided by movements such as muscular pumping

Acuminate – Tapering to a long point

Adecticous – A type of pupa in which the mandibles are immovable and nonfunctional

Adel-, Adelo – Prefix and root-word from Greek: *adelos*; obscure

Adelognatha – A group of weevils (curculionidae) having a short rostrum and temporary mandibles which are later cast off. The larvae usually live in soil, feeding on roots.

Adephaga – One of the suborders of coleopteran; the other being the polyphaga. Adephaga usually have filiform antennae and five-jointed clawed tarsi.

Aedeagus – The male intromittent organ or 'penis' in certain insects.

Aesthetascs – Sensory or olfactory hairs on the antennae or antennules of many orthropods

Aestivation – A state of inactivity which some animals undergo during long periods of drought or heat

Age distribution – The proportion of individuals in different age groups of a population at any given time

Agroecosytem – Man-made crop ecosystem to maximize the production of an agricultural commodity

Airsac – A pouch-like expansion of trachea

Alarm – Pheromone; a interspecific chemical messenger produced to elicit aggressive or defective behaviour

Alate – Winged

Alarypolymorphism – Morphism due to intraspecific diversity in wing

Alienicolae – Wingless female aphids that reproduce parthenogenetically in enormous numbers on herbaceous plants throughout the spring

Alimentary canal – The tube of the digestive system, extending from the mouth to the anus including the foregut, midgut and hindgut

Alinotum – Wing bearing notal plate of the mesothorax or metathorax

Allelochemics – Chemicals involved in communication between members of different species

Alloiogenesis – Alternate sexual and asexual reproduction

Allometic growth – Differential rate of growth whereby the sizes of certain parts of the body are constant exponential function of the size of the whole animal. In male stag- bettles the mandibles become relatively much larger in proportion to the body.

Allomone – An interspecific chemical messenger that benefits the releaser but not the receiver e.g. repellents, toxicants

Allosematic protection – A method of protection in which an animal regularly associates itself with another which is poisonous, distasteful or dangerous

Altruistic behaviour – Self sacrifice

Ambullateral – Walking

Ametabolous – Development without change of form

Ammoniotelic – Aquatic insects excreting ammonia as nitrogenous waste

Amphineustic – A type of insect respiration with mesothoracic and postabdominal functional spiracle

Amphitoky – Production of both males and females parthenogenetically

Anal appendages – Movable appendages at the end of the abdomen

Anal area of the wing – The posterior portion of the wing which usually include the veins

Anal cell – A cell in the anal area of the wing

Anal lobe – A lobe in the porterior basal part of the wing

Anal margin – The posterior edge of the insect wing

Androconia – Wing; scales and connected scent glands of male butterflies
Anecdysis – A long passive period between two moults of an arthropod
Anemotaxies – Directed response relative to an air current
Anepimeron – The upper part of the epimeron of an insect
Anepisternum – The upper part of the epimeron of an insect
Anteapical – Just proximad of the apex
Anteclypeus – An anterior division of the clypeus
Antecasta (*pl* antecostae) – An internal ridge on the anterior portion of a tergum or sternum that serves as the site of attachment of the longitudinal muscles
Antecostal sclerite – A sclerite of the mesosternum, just anterior to the hind coxae
Antecostal suture – An external groove that marks the position of the internal antecosta
Antenna (*pl* antennae) – One of a pair of segmented sensory organs located on the head above the mouth parts
Antennal club – The enlarged distal segments of a clubbed antennae
Antennal fossa – A cavity or depression in which the antennae are located
Antennal groove – A groove in the head capsule in which the basal segment of the antenna fits
Antennal lobes – Lobes in the insect brain; they receive both the sensory and the motor axon from the antennae
Antennal sutures – A suture that surrounds the antennal base and delineates the antennal sclerite
Anus – The posterior opening of the alimentary canal
Aphrodisia – A chemical substance that stimulates sexual desire
Apneustic – The closed types of ventilatory system without functional spiracle
Apodous – Without legs
Aporous – Without pore
Aposematic – Having a warning colouration or repellent smell
Appendage – An external limb or similar structure arising from segments of head thorax or abdomen
Apterous – Wingless
Apterygota – A subclass of primitive wingless insects
Aquatic – Living in water
Arachnida – The arthropod class that includes the spiders, scorpions, ticks and mites
Aranae – An alternative name for Araneida; spiders
Arista – A bristle-like appendage on the antenna of an insect
Aristate – Having aristae on the antennae
Aristopedia – Insect in which the antennae are thickened and leg-like
Arbium – A median pad between the claws of an insect's foot
Arthro – Prefix from Greek *Arthron*: a joint

Arthropod – An irvertebrate animal with jointed appendages; a member of the phylum arthropods

Articul-, Articuls – Prefix and root from Latin *Articulus*: a joint diminutive of Artus

Artophied – Reduced in size, rudimentary

Asco – Prefix from Greek *Askas*; a wine skin

Asexual – Without sex

Aspid, **Aspido** – Prefix from Greek *Aspin, aspidos*: a shield

Asymmetrical – Not alike on the two sides

Attenuated – Very slender and gradually tapering distally

Autotomy – The voluntary casting off a part of the body when an insect is attacked e.g. nymphs of walking stick

Axillary sclerities – The small sclerites at the base of the wing that translate deformation of the thorax into wing movement

Basal areole – A small cell at the base of the wing

Basal cell – A cell near the base of the wing, bordered at least in part by the unbranched portion of the longitudinal vein

Batesian mimic – A palatable and otherwise harmless species that gains protection from predators by virtue of its close resemblance to an unpalatable or harmful species

Bilateral symmetry – A type of body organization in which various parts are arranged more or less symmetrically on either side of a median vertical plane

Binomial – Having two names

Bio – Prefix from Greek *Bios*: life

Biogenesis – The generalization that living organisms must come from parents similar to themselves. In former times the idea of spontaneous generation from non-living matter was held.

Biological control (Biocontrol) – The employment of any biological agent for the control of pest

Biological species – A group of interbreeding organism

Biomass – The total dry weight or volume of living organism in a given area

Biotic potential – A measure of the innate ability of population to survive and reproduce

Biotype–A biological strain of an organism, morphologically indistinguishable from other members of its species, but exhibiting physiological characteristics

Bisexual – With males and females

Brachypterous – Having very short wings

Broad – Several cohorts of offspring produced by a parents or parent population at different times or in different places

Buccal cavity – The cavity of the mouth usually containing the teeth and tongue

Buccal glands – Any glands e.g. salivary, containing the teeth and tongue

Caecum (*pl* **caeca**) – A sac-like or tube-like structure open at only one end e.g. gastric caecum

Calcaria – Movable spurs at the apex of the tibia

Campaniform receptors – Sense organs of insects, probably chemo sensitive, in the form of bell-like projections from the skin with bipolar nerve cells beneath them

Campodeiform larvae – Oligopod larvae; the common form of larvae among beetles

Cannibalism – The behaviour of feeding on other individuals of the same species

Cantharidae – Carnivorous beetles which digest slugs and snails etc.

Cantharidin – An irritating substance produced by blister beetles; capitates with an apical knob-like arrangement e.g. Capitates antenna

Carabidae – Ground beetle

Carbiform larva – A larva resembling carabid larva i.e. elongate, flattened, with well developed legs and antennae and usually active

Cardo (*pl* **cardines**) – Latin cardo: a hinge; any hinge-like part of an organ e.g. the basal segment of the maxilla of an insect

Carina (*pl* **carinae**) – A ridge or keel

Carotenoids – Insect pigments of red and yellow derived from plant tissue

Carrying capacity (k) – The maximum density of a population of a given species that an ecosystem can sustain

Caterpillar – The larva of a butterfly, moth, sawfly or scorpion fly

Caudal – Appertaining to the tail

Cecidomyidae – Gall midges

Cephalo thorax – The fused head and thorax of many crustaceans and certain arthropods

Cercus (*pl* **cerci**) – One of a pair of appendages at the posterior end of abdomen

Cervical groove – A transverse groove indicating the line of demarcation between head and thorax

Cervicum –The neck-region of an insect or other arthropod

Cervix – The neck or any neck-like structure

Chaet, chaeto – Prefix from Greek *chaite*: hair

Chaetotaxy – The arrangement of cuticular structures e.g. hairs, setae etc.

Chel – Prefix and root from Greek *Chele*: a claw

Chemoreceptor – A sensillum capable of detecting chemicals either by olfaction or gestation or by both

Chemotaxis – Reaction to chemical stimuli e.g. smell and taste

Chitin – A horny protective substance forming the cuticle of insects and other arthropods. It is an amino-polysaccharide and resist most solvent agents.

Chitosan – The polysaccharide derived from chitin

Chrysalis – The pupa of a butterfly

Class – A subdivision of a phylum or subphylum containing a group of related orders

Clavate – Club-shaped or elongated at the tip

Claw – A hollow, sharp, multicellular organ, generally paired, at the end of insect leg

Cleptoparasitism – A process in which members of one species take over the nest of another species and feed on the food stored for the host larvae.

Cleridae – Ant-beetles; brightly coloured predatory beetles commonly found feeding on the smaller bark-beetles which infest pine trees

Clubbed – With the distal part enlarged, e.g. clubbed antennae

Clypeus – A cuticular plate of an insect's head immediately anterior to the forns and above the labrum

Cocoon – Case of silk or silk bound debris in which a pupa develops

Coeloconic receptors – Olfactory or other sense organs of an insect

Coelomic sacs – Small cavities in the mesoderm of a developing insect

Coleoptera – Order (coles sheath, ptera, wing) e.g. beetles, weevils etc.

Commensalism – A symbiotic association in which one individual benefits and the other is neither helped nor harmed

Compound eye – The primary visual sensory organs which are located on the either side of insect head

Concave and convex veins – A vein protruding from the lower surface of the wing

Condyle – A knob-like process forming an articulation

Conjunctiva – Soft membranous area separating selerites in the insect skeleton

Copulation – The joining of male and female genital structure

Costa – A longitudinal wing vein usually forming the anterior margin

Costal margin – Anterior edge of the wing

Coxa – The basal segment of the leg

Coxopodite – The basal segment of an arthropod appendage

Crawler – Newly emerged active immature form of a scale insect

Creeping locomotion – The peristaltic movement of body by contraction in the direction of locomotion in legless larvae and caterpillars

Cuticle – The non-cellular outermost layer of insect integument secreted by the epidermis

Cust – A sac, vesicle or bladder like structure

Dealation – Loss of wings by ants or other insects

Decticous – A type of pupa with movable, functional mandibles

Decumbent – Bent downward

Dentate – Toothed or with tooth-like projection

Depress – To lower an appendage e.g. leg or wing

Diapause – A physiological state of arrested metabolism, growth and development that occurs in the life cycle to overcome diverse environmental conditions

Dicyclical – Having two sexual cycle a year, usually one in the spring and the other in the autumn

Dioecious – Having the male and female organs in different individual duals, any one individual being either male or female

Diurnal – Active during day time

Divergent – Becoming more separated distally

Diversicorina – An ill defined group of beetles belonging to the suborder polyphaga.

Dormancy – Suppression of activity associated directly with changes in the abiotic environment

Dorsal – Top or uppermost, pertaining to the back or upper side

Dorsoventral – Form the upper to lower surface or top to bottom e.g. dorsoventral muscle

Ecdyses – Moulting, the periodic shedding of the cuticle of an arthropod

Ecology – The study of all the living organisms in an area and their physical environment

Ecosystem – A system consisting of a living community of plants and animals together with their physical environment of air, water, soil etc.

Ectoparasites – Parasites which live on the outside of their hosts or near the outside

Elbowed antenna – An antenna with the first segment elongated and the remaining segments coming off the first segment at an angle

Elytra – Thickened horny fore-wing of beetles etc.

Emarginate – Notched or indented

Emergence – The act of the adult insect leaving the pupal case or the last nymphal skin

Endopterygota – Insect which passes through a complete metamorphosis in which the larva is very different from the adult and in which there is always a pupal stage

Endoskeleton – A skeleton or supporting structure on the inside of the body

Entognathous – Having mouthparts somewhat pulled or fully withdrawn into the head

Entero – Prefix from Greek *Enteron*: intestine

Enteron – The alimentary canal

Epicranial suture – Coronal plus frontal sutures

Epiculticle (or **Epicuticula**) – Unpigmented outer layer of the insect cuticle

Epipharynx – The membranous roof of the mouth of an insect, provided with taste receptors

Episternum (*pl* **Episterna**) – Anterior portion of thoracic pleuron

Epistomal suture (or **Sulcus**) – The sulcus between the fronts and clypeus that connects the anterior tentorial pits

Epistome – The part of the face just above the mouth; the oral margin

Epiproct – A small tergal plate above the anus of an insect

Eruciform larva – A name formerly used for a polypod or caterpillar type of larva

Exarate pupae – Pupa with free appendages usually not covered by a cocoon
Excavated – Hollowed out
Excretion – Elimination of metabolic wastes from the body
Exopterygotes – Refers to insects in which the wings become externally apparent in the immature instars
Exorted – Protruding or projecting from the body
Exoskeleton – A skeleton or supporting structure on the outside of the body
Exuvae – The cast-off skin of an insect etc.
Exuvial fluid – A mixture of enzymes secreted by the epidermal cells that contain chitinase and protease and digest old endocuticle
Facultative parasite – An organism which is sometimes free-living and sometime parasitic
Faeces – Waste matter egested from the anus
False legs – The prolegs
Family – A subdivision of an order, suborder or superfamily
Fauna – All the animals in a particular locality or habitat
Fecundity – The rate at which females produce egg (ova)
Femur (*pl* Femora) – The third long segment of the leg found between the trochanter and the tibia
Fenestrae – Opening in the dorsal diaphragm through which haemolymph can pass
Flabellate – With fan-like process or projection e.g. Flabellate antennae
Flegellum (*pl* Flagella) – The long, apical antennal segments beyond the scape and pedicel
Flavines – Greenish yellowish pigments found in insects
Foliaceous – Leaf like
Foregut – The anterior portion of the alimentary tract from the mouth to the midgut, also called the stomodeum
Forssorial – Legs bearing sclerotized claw for digging
Frass – Plant fragments, usually of wood boring insects mixed with insect excrement
Front – The portion of the head between the antennae, eyes and ocelli
Frontal lunule – A small crescent-shaped sclerite located just above the base of the antennae and below the frontal suture
Frontal sutures – Two sutures dividing ventrally across the anterior portion of the head
Galea – Jongue-like, distal lobe attached to the stripes of the maxilla
Galeo – Prefix from Latin *Galea*: a helmet
Gamete – A male or female reproductive cell
Gaster – The abdominal
Genae – The lateral cuticular plates of an insect's head, below and behind eyes
Genus (*pl* genera) – A group of closely related species, the name in bionomial or trinomial scientific name

Gill – A respiratory organ in the aquatic immature stage of many insects that obtains dissolved oxygen from water

Glabrous – Smooth, without hairs

Gland – An organ which secretes specific chemical

Glossae – A pair of tongue-like lobes in the middle of the labium of an insect

Glossae (Sin Glassa) – Inner distal lobes found in prementum of the labium

Glottis – The opening from the back of the pharynx into traches

Gnatho – Prefix from Greek *Gnathes*: a jaw

Gula – A sclerite on the ventral side of head between the labium and the formen magnum

Gut – The alimentary canal

Haemocoel – Body cavity of the insect in which viscera are located and the haemolymph circulates

Hatching – The breaking of the egg shell by an insect in the process of eclosion

Haustellate – Having mouthparts adopted for sucking activities

Hemi – Prefix from Greek *Hemi*: half

Hemielyton (*pl* Hemielytra) – The forewing of the tree bugs (heteroptera) the basal portion of which thickened and apical portion membranous.

Hemimetabola – Insects which do not undergo a complete metamorphosis

Hemipneustic – A term used to denote insects with one or more pairs of spiracles

Herbivorous – Feeding on plants

Hibernation – Winter dormancy

Hind gut – Posterior region of the insect alimentary canal proctodaeum

Holopneustic – Pertaining to open ventilatory system that consists of two lateral rows of 10 spiracles each

Holotype – The single specimen selected by the author of a species as its type or the only specimen known at the time of description

Homodynamic life cycle – A life cycle in which there is continous development, without a period of dormancy

Homogenous – Uniform in structure or composition

Homoneura – Moths having almost identical venation in the two pairs of wing

Hypermetabola – The group of insects having several distinct and different forms of larva

Hypermetamorphosis – A type of complete metamorphosis in which there are two or more distinct forms of larvae

Hypognathous – Mouthparts hung vertically from the head capsule

Hypopharynx – A median, tongue-like mouthpart structure anterior to the labium associated with ducts from the salivary gland

Hypopygium – The terminal segments of the abdomen of an insect, often modified for copulation

Imgo (*pl* **imagoes** or **imagines**) – The adult or reproductive stage of insects

Incubation – The time between the laying of an egg and the hatching of young

Innate behaviour – Genetically more-or-less fixed behaviour

Inquilines – Animals that are residents in the shelters or nests of other species

Insecta – Insects; arthropods which breathe by spiracles and tracheae and have the body divided into a distinct head, thorax and abdomen

Insectoverdins – Green pigments found in insect

Instar – Each development form of an insect's life cycle, the insect between successive ecdyses or between successive apolyses

Integument – An outer protective covering

Intercalary segment – The premandibalar segment of the insect head

Jugur – In certain Lepidoptera and Trichoptera, a basal lobe of forewing which overlaps the hind wing, coupling the wings during flight

Juvenile hormone – A hormone produced by corpora allata that promotes larval development and inhibits development of adult characteristics

Kairomone – An interspecific messenger substance that benefits the receiver but not the releaser e.g. attractants, excitants

Katpisternum – The lower part of the episternum of an insect, the upper part is called anepimeron

Key – A tabular arrangement of species, genera, orders or other classification categories according to characters and traits that serve to identify them

K-strategist – A species characterized by low reproductive rate and high survival rate

Labial gland – Exocrine gland opening on or at the base of the labium, usually functioning as salivary or silk gland

Labial palp – One of a pair of small feeler-like structures arising from the labium

Labial suture – The suture on the labium between the postmentum and prementum

Labium – The lower lip of the mouthparts just posterior to the maxillae

Labrum – The upper lip of the mouthparts lying below the clypeus and in front of the mandibles

Lacinia – Distal, mesal lobe of the stripes of a maxilla that bears teeth on its inner edge

Lamella (*pl* Lamellae) – Prefix from Latin *Lamella*: diminutive of *Lamina*, a thin plate

Lamellicornia – Beetles such as the stag-beetles and the chafors which have the distal end of each antenna expanded into a flattened comb-like club or lamella

Larva (*pl* larvae) – The immature stage between the egg and pupa of an insect having complete metamorphosis

Larvifom – Larva like

Larviparity – A form of viviparity in which the female gives "birth" to larvae instead of depositing eggs

Life cycle – The chain or sequence of events that occurs during a lifetime of an individual organism
Life table – A tabulation that accounts for age specific death in population
Ligula – The terminal lobe of labium, the glossae and paraglossae
Lingua – The tongue or any analogous organ e.g. the hypopharynx of an insect
Longitudinal veins – Wing supporting structures with a basal to apical orientation
Luminescence (Bisluminescence) – The production of cold light by certain insect
Lyotidae – Powder post beetles
Macropterous – Large winged
Mandible – A pair of highly sclerolized, unsegmented "jaws" located between the labrum and maxillae, stylets in many insects with piercing-sucking mouthparts
Mandibulate – Having primitive, usually chewing mouthparts
Mandibular palps – Palps or feelers borne on the mandible of some arthropods
Mask – The elongate labium which can be thrust forward to capture prey e.g. Dragonfly naiad
Maxilla – Paired, segmented, secondary jaws that aid in holding and chewing foods
Maxillary palp – A small feeler-like structure arising from the maxilla
Media – The longitudinal vein between the radius and cubitus
Medial cross vein – A cross vein connecting two branches of media
Meloidae – A family which includes oil beetles and Blister beetles
Membranous wings – The forewings or hind wings or both are membranous and supported by tubular veins
Meseepisternum – The episternum of the mesothorax
Mesenteron – The midgut, primary site of digestion and absorption
Mescpimeron (*pl* mescpimera) – The epimeron of the mesothorax
Meso – Middle
Mesocutide – A procuticular layer between the exocutide and endocuticle
Mesonotum – Dorsal plate of the second thoracic segment
Mesoscutum – The scutum of the mesothorax
Mesosternum – The sternum or ventral sclerite of the mesothorax
Mesothorax – Middle segment of the thorax
Metabola – Insect which undergo metamorphosis
Metamorphosis – The change in morphology from which most insect undergo as they develop
Metathorax – The third or posterior segments of the thorax
Metepimeron (*pl* metepimera) – The epimeron of the metathorax
Metepisternum (*pl* metepisterna) – The episternum of the metathorax
Microtrichia – Minute fixed hairs that lack the basal articulation characteristic of seate

Midgut – Middle region of the insect alimentary canal
Mimicry – Protective resemblance of one species to another
Monaliform – Bead-like, with rounded segments
Monogamy – Mating with a single individual
Monophagus – Feeding on one type of food
Morphogenesis – Evolution of form and structure
Moulting – The process of digesting the old cuticle, secreting the new cuticle and shedding the old cuticle
Multivoltine – Having more than one generation per year
Mutation – Something qualitatively new appearing abruptly without transitional form
Mutualsim – A symbiotic relationship between two different species in which both jointly benefit; usually obligatory
Naid – Aquatic immature instar of hemimetabolous insects
Natality – Rate of birth
Natatorial – Legs adopted for swimming
Natural enemies – Living organisms found in nature that kill insects outright, weaken them or reduce reproductive potential
Necorophoridae – Burying beetle
Neoptera – The infraclass of pterygota comprising insects capable of wing flexion
Niche – The ecological position which best suits a particular plant or animal
Nitidulidae – Small predatory beetles often inhabiting the tunnels of bark beetles and weevils and feeding on their eggs and larvae
Notal wing process – Point at which the notum articulates with the wing or axillary sclerites at the base of the wing
Notum – The dorsal sclerite of the thoracic segment of the fused second gonaphyses of the ovipositor
Nymphs – Immature instars of ametabolous and hemimetabolous insects
Obligate parasites – Organisms that require a host to survive
Occipital suture (or **sulcus**) – A line that runs from the posterior end of the coronal suture to just above the mandibles on either side of the head
Occiput – The posterior head region located between the occipital and postoccipital suture
Ocellus (*pl* ocelli) – The small, simple eye of an insect or other arthropods
Oesophagous – The part of the alimentary canal leading from the pharynx to stomach
Omnivorous – Feeding on variety of foods
Oxychophora – A small subphylum of primitive arthropod
Operculum (*pl* opercula) – A lid or cover
Opisthognathous mouthparts – Mouthparts directed ventro-posteriorly relative to the head capsule
Order – A category of classification

Ortho – Prefix from Greek *orthos*: straight
Orthoptera – Order (Ortho: Straight; ptera: wings) e.g. Grasshoppers, crickets and relatives
Oviposition – The act of egg laying
Ovoviviparous – Producing active young ones immediately following the hatching of the eggs within the female
Paedogenesis – Reproduction by immature insects
Palpifer – A lobe bearing a palp, e.g. on the maxillae of insect
Palpiger – The lobe of the mentum of the labium that bears the palp
Palpus (*pl* palp) – A sensory organ associated with mouthpart structures
Paraglossa (*pl* paraglossae) – Two outer lobes found on the prementum of labium
Parasitoid – An orthropod that parasitizes and kills as arthropod host and it is parasitic in its immature stages but is free living as adult
Parthenogenesis – Production of individuals from unfertilized eggs, asexual reproduction
Patyform larva – An extremely flattened larva e.g. aquatic beetle
Paurometabolus – Undergoing gradual metamorphosis passing insects
Pedicel – The second segment of the antenna of an insect
Peripneustic – The respiratory system of an insect in which lateral spiracles occur on all the segments of abdomen
Pest – A species that interferes with human activities property or healthy or is objectionable
Petiolate – Attached by a narrow stalk or stem
Pharynx – The anterior portion of the foregut between the mouth and the oesophagous
Phototaxis – Directed response relative to light
Phylogeny – The origin and evolutionary line of any organism or its evolutionary relationship with other organisms
Phylum – One of the larger groups used in classifying plant and animals
Phyto – Plant
Phytophagous – Herbivorous; feeding on plants
Phytotoxic – Injure or kill plants
Polose – Covered with hair
Placoid sensilla – Plate-like, cuticular sensory structure
Plantulae – Adhesive pods on the tarsal sclerities of some insects
Pleural wing process – Ventral articulation of wing
Pleuron (*pl* pleura) – A lateral plate of a thoracic segment
Plumose – Feather-like e.g. plumose antenna
Plerergate – A worker ant with stomach enormously distended with food
Polygoneutic – Having many broods in a season
Polymorphism – Many forms, for example, one or both sexes of a single species may occur in two or more clearly distinct form
Polyphagous – Feeding on many types of food

Population – A group of individuals of same species within given space and time constrains

Postgena (*pl* postegenae) – A sclerite on posterior lateral surface of the head, posterior to the gena "cheek"

Postmentum – The proximal part of the labium of an insect, usually divided into two parts known respectively as the mentum and submentum

Postnotum (*pl* postnota) – A notal plate behind the scutellum bearing a phragma, often present in wing bearing segment

Predaceous – Feeding as a predator

Predator– An organism that attacks and feeds on other animals and consumes more than animal in its lifetime

Prefemur – The second trochanter segment of the leg

Pregenital – Anterior to the genital segments of the abdomen

Prelabium – The distal part of the labium, comprising the prementum, the ligula and the palpi

Prementum – Apical portion of the labium that bears the labial palpi

Preoral cavity – Cavity formed by the mouthparts

Preoviposition period – The period of time between the emergence of adult female insect and the start of its egg laying

Prepupa – A quiescent stage between the larval period and the pupal period

Prescutum – Anterior sclerite divided off from the alinotum of a wing-bearing segment

Presternal suture – The line that separates the presternum from the rest of the sternum

Presternum – A sclerite on the anterior part of sternum divided off by the presternal suture

Pretarsus (*pl* pretarsi) – The terminal segment of the leg, usually bearing a pair of claws and one or more pod-like structures

Prey – Organisms eaten by predators

Proboscis – The extended beak-like mouthparts of sucking insects

Proclinate – Inclined forward or downward

Procuticle – Endocuticular and exocuticular layers together

Prognathous – Having a forward projecting face and jaws as in arthropod apes

Prolegs – Short unjointed limbs of some of the abdominal segments of caterpillars

Pronotal comb – A row of strong spines borne on the posterior margin of the pronotum as in Siphanopter

Pronotum (*pl* pronota) – Dorsal plate of first thoracic segment

Propleuron – One of the lateral parts of the prothorax

Prosternal grooves – A groove in the sternum or the ventral sclerite of the prothorax

Prosternum – The sternum or ventral sclerite of the prothorax

Prothorax – The first or anterior segment of the thorax

Proximal – Adjacent to or nearest the point of attachment or origin .
Pseudo – False, temporary
Pseudopupa – A coarctate larva; a larva in a quiescent pupal-like condition, one or two instars before the pupal stage
Ptero – Prefix from Greek *Pteron*: a wing
Pterostima – A thickened opaque spot along the costal margin of the wing near the wingtip as in Odonata, also called the stigma
Pteregate – An exceptional form of worker ant having the rudiment of wings
Pterygota – The subclass of insect comprising the wings and secondarily wingless insects
Pterygote – Winged insect, a member of the subclass pterygota
Pubescent – Covered with short fine hairs
Pulvillus – A pod or cushion between the claws of an insect
Pupa – The third stage in the life of an insect
Puparium (*pl* puparia) – A case formed by the hardening of the last larval skin which encloses a pupa as in dipteral
Poparous – Giving birth to larvae that are full grown and ready to pupate
Pygidium – The last dorsal segment of the abdomen
Pygopodia – Accessory locomator appendages that arise from the terminal abdominal segment in some hexapodous larval form
Puadrate – Four sided
Puadrivalline – Four life cycles per year
Pueem – A fertile female
Puiescence – Inactivity in response to adverse environmental conditions that reverses when favourable environmental conditions return
Race – A variety of a species with constant characters
Radius – The longitudinal vein between the subcosta and the media
Rank – The level in a taxonomic hierarchy e.g. genus, family, order and so on
Raptorial (legs) – Forelegs modified for grabbing and holding prey as in praying mantids
Reclinate – Included backward or upward
Regression – The amount of change in one variable associated with a unit change in another variable
Reticulate – Like a network
Rhin-, **Rhino** – Prefix from Greek *Rhis*, *rhinos*: nose
Rhizo – Prefix from Greek *Rhizo*: root
Rudimentary – Reduced in size
Ruguse – Wrinkled
Sacculus – Any small sac or vesicle
Salivarium – The cavity between the hypopharynx
Salivary glands – Exocrine glands typically associated with the labial segment, highly variable in size, structure and function
Saprophagous – Feeding on dead and decaying organic matter
Sarothrum – The pollen brush on the hind leg of bee

Scape – Basal segment of an insect antenna

Scarabaciform larva – A grub-like and usually sluggish larva with a well developed head and thoracic legs thickened and cylindrical body and without prologs

Scent gland – A gland producing an odorous substance

Scutellum – The hindmost part of the notum of an insect, behind the scutum

Semiochemical – Any chemical involved in communication among organisms

Serrate – Toothed along the edge like a saw e.g. serrate antennae

Setaceous – Bristle-like e.g. setaceous antennae

Sigmoid – Shaped like the letter S

Silphidae – Burying and carrion beetle

Species – A group of animals or plants having a high degree of similarity, able to breed among themselves but not generally to breed with members of another species

Spine – A multicellular more or less thorn-like process or outgrowth of the cuticle

Spiracle – An external opening of the tracheal system; a breathing pore

Spurs – Thick cuticular hairs or spines on the legs of certain insects

Staphylinidae – Rove beetle; a large family of beetles, some carnivorous and some herbivorous, all characterized by the short elytra which leave the abdomen exposed

Stemmata (Sing Stemma) – Simple eyes found on the sides of the insect head when compound eyes are lacking

Sternites – Sclerites that form subdivisions of the abdominal sternum

Sternum (*pl* sterna) – Ventral region of a thoracic or abdominal segment

Sting – A defensive or offensive, organ usually able to pierce the prey

Stomodaeum – The foregut, beginning with mouth and ending with proventriculus

Strain (*pl* **striae**) – A grove or depressed line

Striate – With grooves or depressed line

Stylus (Style, Stylet) – Any pointed bristle-like appendage

Subalar sclerite (**Subalare**) – Small sclerite posterior to the pleural wing process

Subantennal sulcus – A groove on the face extending ventrally from the base of the antenna

Subclass – A major subdivision of class containing a group of related suborder

Subcosta – The longitudinal vein between the costa and the radius

Subcoxa – Leg segment of primitive arthropods

Subcuticle – Newly secreted endocutiele in which the microfibres are not oriented

Subfamily – A taxonomic subdivision of a family containing a group of closely allied genera different from other allied groups

Submentum – The basal segment of the labium of an insect

Subneural – Beneath the nerve cord

Subocular sutures – Vertical lines that may run beneath the compound eye
Suborder – A major subdivision of an order containing a group of superfamilies
Subspecies – A subdivision of a species, usually a geographic race
Successions – Group of species that successively occupy a given habitat as the conditions of the habitat change
Sucking pump – The highly modified cibrarium, in insects which ingest liquid food
Sulcus – Any groove or fissure, particularly those in cerebral hemispheres
Superlinguae – Small paired lobes fused to the sides of hypopharynx in certain insects
Sutures – External seams (grooves) or lines in the insect skeleton indicating the division of distinct parts of the body wall, also called sulci
Swarm – A concerted departure or association of a large group of insects
Symbiont – An organism living in symbiosis with other organism
Symbiosis – A close association between two or more different species that benefits one or both of the species
Tagmata (Sing tagma) – The three major regions of the insect body: head, thorax and abdomen
Tarsal claw – A claw at the apex of the tarsus, derived from the pretarsal segment of the leg
Tarsomere – A subdivision or "segment" of the tarsus
Tarsus (*pl* tarsi) – Distal segmented part of insect leg attached to the tibia
Taxon (*pl* taxa) – A group of organisms classified together
Tegmen (*pl* tegmina) – The thickened or leathery forewing of an Orthopteran insect
Telum – The last abdominal segment
Tenet hairs – Tiny hairs on the pulvilli and tarsal pods that allow some insects to cling to smooth surface
Tentorium – The endoskeleton of the head of an insect consisting of a number of chitinous struts connecting the exoskeleton to a transverse bers in the centre of head
Tergum (Tergite) – A dorsal cuticular plate covering each body segment of an arthropod
Tetra – Prefix from Greek *Tetrva*: four
Tetrapterous – Four winged
Tysanoptera – Thrips; minute insects with short antennae and narrow wings fringed with hair
Tibia (*pl* tibiae) – The fourth long segment of the leg between the femur and the tarsus
Tibial super – A large spine on the tibia, usually located at the distal end of the tibia
Tribe – A subdivision of a subfamily containing a group of related genera
Trichogen cell – The epidermal cell from which a seta develops
Trochantin – A small basal articular sclerite on the trachanter of an insect

Tympanic organs – Auditary organs composed of a thin integrumental area (tympanic membrane) and a group of chordotonal sensila attached to the ended surface

Uniporous – With a single pore

Unisetose – Possessing one bristle or seta

Unisexual – Consisting of or involving only female

Uric acid – The primary nitrogenous waste in insects

Veins – Vessels which convey blood from the various parts of the body back to the heart

Vermiform – Worm-like

aVespidae – The social wasps

Visceral – Pertaining to the internal organs of the body

Viviparous – Giving birth to active young one which undergo growth and development inside the mother

Wax layer – An epicuticular lipid layer between the cuticulin layers that contributes to the permeability characteristics of the cuticle

Wing cells – The areas of the wing enclosed by vein

Wing veins – Longitudinal and transverse support framework of the wing

Xanthophyll – A blood pigment in phytophagous insect

Xenarthra – Ant eaters

Xiph – Prefix and root word from Greek *Xiphos*: a sword

Yolk – The food store of an egg consisting largely of proteins and fats

Yolk nutrients – Carbohydrates, proteins and lipids found in mature insect eggs; also called the deutoplasma

Zoo – Prefix from Greek *Zoon*: an animal

Zoophagous – Feeding on animals

Zoraptera – Order (Zor: pure, aptera: wingless)

Zyg-, **Zygo-** – Prefix from Greek *Zugon*: a yoke usually signifies anything jointed together

Zygote – A fertilized ovum; a cell formed by the union of male and female gamete

Appendix D

Standard Abbreviation

Units and General Abbreviations

a.i Active ingredient
°C Degrees Celsius
cm Centimetre
e.c. Emulsifable concentrate
g Gram
h Hour
ha Hectare
h.v. High volume
i.r. Infra red
kg Kilogram
km Kilometre
l Litre
LC_{50} Medial lethal concentration
LD_{50} Median lethal dose
l.v. Low volume
m Metre
mg Milligram
min Minute
ml Millilitre
mm Millimetre
NMDS Non-metric multidimensional scaling
pH Hydrogen ion concentration
ppm Parts per million
RH Relative humidity
S Second
Sp. Species
Spp. species (plural)
Ssp Subspecies
Sspp Subspecies (plural)
u.l.v. Ultra low volume
u.v. Ultra violet

var. Variety
vol. Volume
w.p. Wettable powder
w/w Weight for weight

Miscellaneous abbreviations

BPH Brown planthopper of rice
BC Biological control
BSI British Standard Institute
Cda Controlled droplet application
GV Granulosis virus
IPM Integrated pest management
EAG Electro antenna-gram
HMSO Her majesty's stationary office, UK
OC Organochlorine compounds
OP Organophosphorous compounds
PHV Polyhedrosis virus
PM Pest management
SIRM Sterile insect release method

General Bibliography

Abhitha, P. and Sabu, T.K. 2009. Rare ground-beetle species of *Leleuporella* Basilewsky (Coleoptera: Carabidae: Scaritinae: Scaritini) from Indian sub-continent. *Zootaxa*, **2310:** 59–63.

Adams, C.C. 1909. An Ecological Survey of Isle Royale, Lake Superior. Wynkoop Hallenbeck Crawford Co., Lansing, Michigan. 468 p.

Adams, C.C. 1915. An ecological study of prairie and forest invertebrates. *Illinois State Laboratory of Natural History Bulletin*, **11:** 33–280.

Albrecht, M., Schmid, B., Obrist, M.K., Schüpbach, B., Kleijn, D. and Duelli, P. 2010. Effect of ecological compensation meadows on arthropod diversity in adjacent intensively managed grassland. *Biological Conservation*, **143:** 642–649.

Alcock, J. 1976. The behaviour of the seed-collecting larvae of a carabid beetle (Coleoptera). *Journal of Natural History*, **10:** 367–375.

Alexander, K.N.A. 1993. Observations from an Irish mountain summit. *Coleopterist*, **2:** 42–44.

Alford, D.V. 2002. Biocontrol of Oilseed and Rape Pests. Wiley. 343 p.

Allen, A.A. 1950a. A second English capture of *Trechus subnotatus* Dej. (Col., Carabidae). *Entomologists' Monthly Magazine*, **86:** 38.

Allen, A.A. 1950b. Two species of Carabidae (Col.) new to Britain. *Entomologists' Monthly Magazine*, **86:** 89–92.

Allen, A.A. 1956. *Amara cursitans* Zimm. (Col., Carabidae) new to Britain. *Entomologists' Monthly Magazine*, **92:** 215–216.

Allen, A.A. 1958. Notes on the larval feeding habits of *Lebia* species (Carabidae). *Entomologists' Monthly Magazine*, **94**.

Allen, A.A.1965. *Harpalus bonestus* Duft. (Col., Carabidae) confirmed as British. *Entomologists' Monthly Magazine*, **100:** 155–157.

Allen, A.A. 1989. The last British capture of *Scybalicus oblongiusculus* Dej. (Col.: Carabidae)? *Entomologists' Record and Journal of Variation*, **101:** 108.

Allen, A.A. 1991. On the separation of *Patrobus atrorufus* Ström and *P. assimilis* Chaud. (Col., Carabidae). *Entomologist's Record and Journal of Variation*, **103:** 71–72.

Anderson, R. 1981. Coleoptera from the Burren, including *Acupalpus consputus* (Dufts.) and *Bembidion fumigatum* (Dufts.) (Carabidae) new to Ireland. *Entomologists' Monthly Magazine*, **116:** 138.

Anderson, J.M. 1972. Food and feeding of *Notiophilus biguttatus* F. (Coleoptera: Carabidae). *Revue d'Ecologie et de Biologie du Sol*, **9:** 177–184.

Andersen, J. and Skorping, A. 1991. Parasites of carabid beetles: Prevalence depends on habitat selection of the host. *Canadian Journal of Zoology*, **68:** 1216-1220.

Anderson, R. 1993. Rufmism in Irish *Pterostichus nigrita* (Paykull) (Col., Carabidae). *Entomologist's Monthly Magazine*, **129:** 122.

Anderson, R. and Luff, M.L. 1994. *Calathus cinctus* Motschulsky, a species of the *Calathus melanocep*/mollis complex (Col., Carabidae) in the British Isles. *Entomologist's Monthly Magazine*, **130:** 131–135.

Anderson, R., Nash, R. and O'Connor, J.P. 1997. Irish Coleoptera: A revised and annotated list. *Irish Naturalists Journal Special Entomological Supplement*, 1–81.

Andrewes, H.E. 1929. The Fauna of British India including Ceylon and Burma, Coleoptera, Carabidae, Vol. I. Carabinae. Taylor and Francis, London.

Anjum-Zubair, M., Martin, H.E., Alexander, B., Thomas, D. and Thomas, F. 2015. Differentiation of spring carabid beetle assemblages between semi-natural habitats and adjoining winter wheat. *Agricultural and Forest Entomology*, DOI: 10.1111/afe.12115. (Accessed on August 14, 2017)

Anlas, S. and Tezcan, S. 2010. Species composition of Ground Beetles (Carabidae, Coleoptera) collected by hibernation trap-brands in agricultural landscapes, Bozdaglar Mountain of Western Turkey. *Acta Biologica Universitalis Daugavpiliensis*, **10(2):** 193–198.

Arnol'di, L.V., Zherikhin, V.V., Nikritin, L.M. and Ponomarenko, A.G. 1991. Mesozoic Coleoptera. Smithsonian Institution Libraries and the United States National Science Foundation. Washington D.C. 308 p.

Arnol′di, L.V., Zherikhin, V.V., Nikritin, L.M. and Ponomarenko, A.G. 1992. Mesozoic Coleoptera. Oxonian Press, New Delhi. 284 p.

Atty, D.B. 1983. Coleoptera of Gloucestershire. Cheltenham. Privately published.

Atwal, A.S. 1986. Agricultural Pests of India and South-East Asia. Kalyani Publisher, New Delhi. 328 p.

Aukema, B. 1991. Fecundity in relation to wing-morph of three closely related species of the Melanocephalus group of the genus *Calathus* (Coleoptera: Carabidaae). *Oecologia*, **87:** 118–126.

Avgin, S.S. and Emre, I. 2007. A check-list of Nebriini (Coleoptera: Carabidae) from Turkey and species belonging to Nebriini tribe collected from Kahramanmaras and the surrounding province. *International Journal of Natural and Engineering Sciences*, **1:** 35–43.

Ayre, K. 2001. Effect of predator size and temperature on the predation of *Derocerus reticulatum* (Muller) (Mollusca) by carabid beetles. *Journal of Applied Entomolgy*, **125:** 389–395.

Baars, M.A. and van Dijk, T.S. 1984. Population dynamics of two carabid beetles at a dutch heathland. II. Egg production and survival in relation to density. *Journal of Animal Ecology*, **53:** 389–400.

Baehr, M. 2003. On a collection of ground beetles from Gambia (Insecta, Coleoptera, Carabidae). *Entomofauna (Zeitschrift für angewandte Entomologie)*. Brand 24. Heft **28:** 397–424.

Balduf, W.V. 1935. The Bionomics of Entomophagous Coleoptera. John S. Swift, New York, NY. 220 p.

Balkenohl, M. 1999. Revision of the genus *Trilophus* Andrewes from the Oriental region (Coleoptera, Carabidae). *Revue Suisse de Zoologie*, **106:** 429–537.

Balkenohl, M. 2001. Key and catalogue of the Tribe Clivinini from the Oriental Realm, with Revisions of the genera *Thiloclivina* Kult and *Trilophidius* Jeannei (Insecta, Coleoptera, Carabidae, Scarititae, Clivinini), Pensoft Series Faunistica No. 21, Softia-Moscow, Pensoft, 86 p.

Ball, G.E. 1956. Revision of North American species of genus *Helluomorphoides. Proceeding of the Entomological Society Washington*, **59:** 67–91.

Ball, G.E. 1959. A Taxonomic Study of the North American Licinini with notes on the old world species of the genus *Diplocheila brulle* (Coleoptera). *American Entomological Society Memoirs*, **16**: 1–258.

Ball, G.E. 1960. Carabidae (Latreilee, 1810). pp. 55-181. *In:* R.H. Arnett Jr. (ed.). The Beetles of the United States, **4:** 55–181. The Catholic University of America Press, Washington D.C. 210 p.

Ball, G.E. and Anderson, J.N. 1962. The Taxonomy and Speciation of *Pseudophonus*. The Catholic University of American Press, Washington, D.C. XI+94 pp.

Ball, G.E. and Bousquet, Y. 2001. Carabidae Latreille, 1810. pp. 32-132. *In:* R.H. Arnett Jr. and M.C. Thomas (eds). American Beetles. Vol. 1. CRC Press. Boca Raton, Florida. xv+443 pp.

Ball, G.E. and Erwin, T.L. 1982. The Baron Maxmilien De Chaudoir-inheritance, Associates, Travels, work and legacy. *The Coleopterists Bulletin*, **36(3):** 475–501.

Bambaradeniya, C.N.B. 2016. The Fauna of Sri Lanka, Status of Taxonmy Research and Conservation. *The World Conservation Union* (IUCN), Sri Lanka. viii+308 pp.

Banninger, M. 1937-1939. Monographie der Subtribus Scartina (Col. Carab.) 1-3. Dt.ent. Zschr.1937: 81–161, 1938: 41–181, 1939: 126–161. Berlin.

Barney, R.J. and Pass, B.C. 1986. Ground beetle (Coleoptera: Carabidae) populations in Kentucky alfalfa and influence of tillage. *Journal of Economic Entomology*, **79**: 511–517.

Basilewsky, P. 1948. Contribution á I'étude des Coléoptéres Carabidae du Congo Belge. II Révision des Tachyini du Congo – *Bulletin du Musée Royal d'Histoire Naturelle de Belgique*, **24(33):** 1–28. Bruxelles.

Basilewsky, P. 1950-51. Révision générale des Harpalides d'Afrique et de Madagascar (Coleoptera, Carabidae, I and II, *Annales du Musee du Congo Belgique Zoologicae*, **6:** 1–283, **9:** 1–333. Bruxelles.

Basilewsky, P. 1950b. Contribution å l'étude de I' Air. Coléoptéres Carabidae – Mem. *Memórias do Institut français d'Afrique noire de*, **10:** 230–260. Dakar.

Basilewsky, P. 1953b. Carabidae (Coleoptera Adephaga). Exploration du. parc national de I' Upemba. *Mission G.F. De Witte.*, **10:** 1–252. Bruxelles.

Basilewsky, P. 1956. Coléoptéres *Carabidae recueillis* par Mr. et Mme. Bechyné en Afrique Occidentale Francaise. *Entomologische Arbeiten aus dem Museum G. Frey Tutzing bei München*, **7:** 439–489. Tutzing.

Basilewsky, P. 1960. Ĕtude des Dryptinae d'Afrique (Coleoptera, Carabidae). *Bulletin & Annales de la Société Entomologique de Belgique*, **96:** 133–182. Bruxelles.

Basilewsky, P. 1961a. Contribution å la connaissance des Coléoptéres des Carabidae de la region du Tschad. I. *Carabides recueillis par le Dr.H. Franz. revue Française d'Entomologie.* **28:** 212–235. Paris.

Basilewsky, P. 1964. Contribution å l'étude des Coléoptéres Carabiques de I' Afrique occidentale. VI. *Bulletin de l'Institut Français d'Afrique Noire*, **26:** 160–175. Dakar.

Basilewsky, P. 1965. Contribution å l'étude de la faune de la basse Casamance (Sénegal). XIII. Coléoptéres Carabidae. *Bulletin de l'Institut Français d'Afrique Noire*, **27:** 204–215. Daker.

Basilewsky, P. 1968. Contributions å la connaissance de la faune entomologique de la Cote d'Ivoire (J. Decelle, 1961-1964). IV. Coleoptera, Carabidae. *Annales du Musée Royal de l'Afrique Centrale de Zoologie*, **165:** 29–124. Bruxelles.

Basilewsky, P. 1969. Le Parc national du Niokol-Koba (Sénégal). 3.21. Coleoptera, Carabidae. *Memórias do Instituto Francais s'Afrique Noire*, **84:** 321–353. Dakar.

Basilewsky, P. 1970. Notes sur les Coléopteres Carabidae de la Mauritanie. *Bulletin & Annales de la Société Entomologique de Belgique*, **106:** 167–174. Bruxelles.

Basilewsky, P. 1980. Les Reicheiina de l'Afrique du Sud (Coleoptera: Carabidae). *Entomologia Generalis*, **6:** 293–302.

Bassett, P. 1978. Damage to winter cereals by *Zabrus tenebrioides* (Goeze) (Coleoptera: Carabidae). *Plant Pathology*, **27:** 48.

Batary, P., Holzschuh, A., Orci, K.M., Samu, F. and Tscharntke, T. 2012. Responses of plant, insect and spider biodiversity to local and landscape scale management intensity in cereal crops and grasslands. *Agriculture Ecosystems and Environment*, **146:** 130–136.

Bauer, T. 1974. Ethologische, autökologische und ökophysiologische Untersuchungen an *Elaphruscupreus* DFT. und *Elaphrus riparius* L. (Coleoptera, Carabidae). *Oecologia* (Berlin), **14:** 139–196.

Bauer, T. and Kredler, M. 1988. Adhesive mouthparts in a ground beetle larva (Coleoptera, Carabidae) *Loricera pilicornis* F. and their function during predation. *Zoologischer, Anzetger*, **221:** 145–156.

Bell, R.T.A. 1960. Revision of the genus *Chlaenius bonelli* (Coleoptera, Carabidae) in North America. *Entomological Society of America Miscellaneous Bulletin*, **1:** 98–166.

Berim, N.G. and Novikov, N.V. 1983. Feeding specialization of ground beetles. *Zasch. Rast.* (Moscow), **(7):** 18.

Best, R.L. and Beegle, C.C. 1977. Food preferences of five species of carabids commonly found in Iowa cornfields. *Environmental Entomology*, **6:** 9–12.

Best, R.L., Beegle, C.C., Owens, J.C. and Ortiz, M. 1981. Population density, dispersion, and dispersal estimates for *Scarites substriatus*, *Pterostichus chalcites* and *Harpalus pensylvanicus* (Carabidae) in an Iowa cornfield. *Environmental Entomology*, **10:** 847–856.

Bhat, P.S. 1984. Faunislic Study of Carabids (Coleoptera: Carabidae) with some Aspects of their Ecology in Bangalore. M. Sc. (Agri) thesis, submitted to University of Agricultural Science, Bangalore. 192 p.

Bigger, J.H. 1934. Notes on the Flight and Abundance of the Seed Corn Beetle *Agonoderus pallipes. Transactions of the Illlinois State Academy of Science*, **26:** 137.

Bilde, T. and Toft, S. 1999. Prey consumption and fecundity of the carabid beetle *Calathus melanocephalus* on diets of three cereal aphids: High consumption rates of low quality prey. *Pedobiologia*, **43:** 422–429.

Birkhofer, K., Wolters, V. and Dickätter, T. 2013. Grassy margins along organically managed cereal fields foster trait diversity and taxonomic distinctness of arthropod communities. *Insect Conservation and Diversity*, **7:** 274–287.

Blatchley, W.S. 1910. Coleoptera, or Beetles Known to Occur in Indiana. *The Nature Publishing Company*, Indianapolis, Indiana. 1386 p.

Blickenstaff, C.C. 1965. Common names of insects approved by the Entomological Society of America. *Entomological Society of America Bulletin*, **11:** 287–320.

den Boer, P.J. 1965. External characters of sibling species *Trechus obtusus* Er. and *T. quadristriatus* Schrk. (Coleoptera). *Tifdschnft voor Entomologie*, **108:** 219–239.

den Boer, P.J. 1970. On the significance of dispersal power for populations of carabid-beetles (Coleoptera, Carabidae). *Oecologia*, **4(1):** 1–28.

den Boer, P.J. 1977. Dispersal power and survival. Carabids in a cultivated countryside. *Miscellaneous Papers*, *Landbouwhoge School Wageningen*, **14:** 1–190.

den Boer, P.J. 1986. What can carabid beetles tell us about dynamics of populations. *In:* P.J. den Boer, M.L. Luff, D. Mossakowski and F. Weber (eds). Carabid Beetles, their Adaptations and Dynamics. Fisher, Stuttgart, Germany.

den Boer, P.J., Huizen, T.H.P. van, Boer–Daanje den, W., Aukema, B. and Bieman, C.F.M. den 1980. Wing polymorphism and dimorphism as stages in an evolutionary process (Coleoptera, Carabidae). *Entomologia Generalis*, **6:** 107–134.

Booij, K. 1994. Diversity patterns in carabid assemblages in relation to crops and farming systems. *In:* K. Desender, M. Dufrêne, M. Loreau, M.L. Luff and J.-P. Mealfait (eds). Carabid Beetles: Ecology and Evolution. Kluwer Academic Publishers, Dordrecht, The Netherlands.

Bonn, A. and Kleinwachter, M. 1999. Relation between brown planthopper, *Nilaparvata lugens* and carabid beetle, *Ophionea indica. Zeitscrift Fur Okologic und Naturschutz*, **8:** 109–123.

Bousquet, Y. 2012. Catalogue of Geadephaga (Coleoptera, Adephaga) of America, north of Mexico. *ZooKeys*, **245:** 1–1722.

Bousquet, Y. and Larochelle, A. 1993. Catalogue of the Geadephaga (Coleoptera: Trachipachidae; Rhysodidae, Carabidae including Cicindelini of America North of Mexico. *The Memoirs of the Entomological Society of Canada*, **125(167):** 3–397.

ter Braak, C.J.F. 1986. Canonical correspondence analysis: A new eigenvector technique for multivariate direct gradient analysis. *Ecology*, **67:** 1167–1169.

ter Braak, C.J.F. 1995. Ordination. *In:* R.H.G. Jongman, C.J.F. ter Braak and O.F.R. van Tongeren (eds). Data Analysis in Community and Landscape Ecology. Cambridge University Press, Cambridge, UK.

ter Braak, C.J.F. and Šmilauer, P. 1998. CANOCO Reference Manual and userÕs guide to canoco for windows: Software for canonical community ordination (version 4.5). Microcomputer Power, Ithaca, NY.

Brandmayr, P. and Zetto-Brandmayr, T. 1979. The evolution of parental care phenomena in Pterostichne ground beetles with special reference to the genera *Abax* and *Molop* (Col. Carabidae). *In:* P.J. den Boer, H.-U. Thiele and F. Weber (eds). On the Evolution of Behavior in Carabid Beetles. *Miscellaneous Paper no. 8. Agricultural University of Wageningen*, **18:** 35–49.

Briggs, J.B. 1957. Some experiments on control of ground beetle damage to strawberry. *Report of East Malling Research Station*, **44:** 142-145.

Briggs, J.B. 1965. Biology of some ground beetles injurious to strawberries. *Bulletin of Entomological Research*, **56:** 79–93.

Britten, H. 1943. The Coleoptera of the Isle of Man (1st part). *North Western Naturalist*, 73–87.

Bruneau De Mire, P. 1963. Les *Tachyini africains* de la collection du Museum National d'Histoire Naturelle de Paris (2nd note). *Rev. fr. Ent.*, **30:** 243–256. Paris.

Bruneau De Mire, P. 1964. Les *Tachyini africains* de la collection du Museum National d'Histire Naturelle de Paris (2nd note), *Rev. fr. Ent.*, **31:** 70–100. Paris.

Bruneau De Mire, P. 1990. Les Coléopéres Carabiquesdu Tibesti. *Annales de la Société Entomologique de France* (N.S.) **26:** 499–544. Paris

Brunsting, A., Siepel, H. and van Schaik Zillesen, P.G. 1986. The role of larvae in the population ecology of Carabidae. *In:* P.J. den Boer, M.L. Luff, D. Mossakowski and F. Weber (eds). Carabid Beetles. Their Adaptation and Dynamics. Fischer Verlag, Stutt/New York. pp. 399-411.

Bryan, K.M. and Wratten, S.D. 1984. The responses of polyphagous predators to prey spatial heterogeneity: Aggregation by Carabid and Staphylinid beetles to their cereal aphid prey. *Ecological Entomology*, **9:** 251–259.

Bousquet, Y. 2010. Review of the Nearctic and West Indian (Greater Antilles) species of *Colliuris* Degeer (Coleoptera: Carabidae: Odacanthini). *Zootaxa*, **2529:** 1–39.

Burakowski, B. 1986. The life cycle and food preference of *Agonum quadripunctatum* (De Geer.) *In:* P.J. den Boer, L. Grum and J. Szyszko (eds). Feeding Behavior and Accessibility of Food for Carabid Beetles. Warsaw Agric. Univ. Press. Warsaw. pp. 35–39.

Burgess, A.F. 1911. *Calosoma sycophanta*: Its life history, behaviour and successful colonization in New England. *United States Department of Agriculture. Bureau of Entomology Bulletin*, **101:** 1-94.

Burgess, A.F. and Collins, C.W. 1917. The genus *Calosoma*: Including studies of seasonal histories, habits and economic importance of American species. *United States Department of Agriculture. Bureau of Entomology Bulletin*, **417:** 1–124.

Burgess, E.P.J., Lovei, G.L., Malone, L., Nielsen, I.W., Gatehouse, H.S. and Chisteller, J.T., 2002. Tri-trophic effects of the protease inhibitor a protinin on the predatory carabid beetle *Nebria bevicollis. Journal of Insect Physiology*, **48:** 1093–1101.

Calow, P. 1979. The cost of reproduction—A physiological approach. *Biological Reviews*, **54:** 23–40.

Cárcamo, H.A. 1995. Effect of tillage on ground beetles (Coleoptera: Carabidae): A farm-scale study in central Alberta. *Canadian Entomologist*, **127:** 631–639.

Cárcamo, H.A., Niemala, J.K. and Spence, J.R. 1995. Farming and ground beetles: Effects of agronomic practice on populations and community structure. *Canadian Entomologist*, **127:** 123–140.

Cardenas, A.M. and Buddle, C.M. (2007). Distribution and potential range expansion of seven introduced ground beetle species (Coleoptera: Carabidae) in Quebec, Canada. *The Coleopterists Bulletin.*, **61:** 135–142.

Cartellieri, M. and Lövei, G.L. 2003. Seasonal dynamics and reproductive phenology of ground beetle (Coleoptera, Carabidae) in fragments of native forest in Manawatu, North Island, New Zealand. *New Zealand Journal of Zoology*, **30:** 31–42.

Casale, A. and Taglianti, A.V. 1999. Carabid beetles (excl. Cicindelidae) of Anatolia, and their biogeographical significance (Coleoptera, Caraboidea).

Biogeographia, Lavoridella Societa Italiana di Biogeografia, Siena, Italy, **20:** 277–406.

Casey, T.L. 1914. A revision of the Nearctic Harpalinae. *Memoirs on the Coleoptera*, **5:** 45–305.

Casey, T.L. 1918a. A review of the North American Bembidiinae. *Memoirs on the Coleoptera*, **8:** 1–223.

Casey, T.L. 1918b. Studies among some of the American Amarinae and Pterostichinae. *Memoirs on the Coleoptera*, **8:** 224–239.

Casey, T.L. 1920. Random studies among the American Caraboidea. *Memoirs on the Coleoptera*, **9:** 133–299.

Chiu, S. 1979. Biological control or the brown planthopper. *In:* Brown planthopper: Threated to Rice Production in Asia. International Rice Research Institute, Los Banos, Philippines.

Chiverton, P.A. 1986. Predator density manipulation and its effects on populations of *Rhopalosiphum padi* (Hom: Aphididae) in spring barley. *Annals of Applied Biology*, **109:** 49–60.

Chiverton, P.A. 1988. Searching behavior and cereal aphid consumption by *Bembidion lampros* and *Pterostichus cupreus* in relation to temperature and prey density. *Entomologia Experimentalis et Applicata*, **47:** 173–182.

Chiverton, P.A. and Sotherton, N.W. 1991. The effects on beneficial arthropods of the exclusion of herbicides from cereal crop edges. *Journal of Applied Ecology*, **28:** 1027–1039.

Choate, P.M. 2003. A Field Guide and Identification Manual for Florida and Eastern U.S. Tiger Beetles. University of Florida Press. 224 p.

Churchfield, J.S., Hollier, J. and Brown, V.K. 1991. The effect of small mammal predators on grassland invertebrates, investigated by field exclosure experiment. *Oikos.*, **60:** 283–296.

Clark, M.S., Luna, J.M., Stone, N.D. and Youngman, R.R. 1994. Generalist predator consumption of armyworm (Lepidoptera: Noctuidae) and effect of predator removal on damage in no-till corn. *Environmental Entomology,* **23:** 617–622.

Cohen, A.C. 1995. Extraoral digestion in predaceous terrestrial Arthropoda. *Annual Review of Entomology*, **40:** 85–103.

Colombini, I., Chelazzi, L. and Scapini, F. 1994. Solar and landscape cues as orientation mechanisms in the beach-dwelling beetle *Eurynebria complanata* (Coleoptera, Carabidae). *Marine Biology*, **118:** 425–432.

Comstock, J.H. 1960. An introduction to Entomology. Comstock Publishing Associate. 1064 p.

Constantine, B. 1993. *Nebria livida* (L.) (Carabidae) at an inland site in east Yorkshire. *Coleopterist*, **2:** 54–55.

Coombes, D.S. and Sotherton, N.W. 1986. The dispersal and distribution of polyphagous predatory Coleoptera in cereals. *Annals of Applied Biology*, **108:** 461–474.

Cornic, J.F. 1973. Etude du regime alimentaire de trios especes de carabiques at de ses variations verger de pomiers. *Annales de la Société Entomologique de France*, **9:** 69–87.

Cromar, H.E., Murphy, S.D. and Swanton, C.J. 1999. Influence of tillage and crop residue on postdispersal predation of weed seeds. *Weed Science*, **47:** 184–194.

Crowson, R.A. 1955. The Natural Classification of the Families of Coleopteran. Nathaniel Lloyd. London.

Crowson, R.A. and Crowson, E.A. 1963. Observations on insects and arachnids from the Scottish south coast. *Glasgow Naturalist*, **18:** 228–232.

Crowson, R.A. 1981. The Biology of the Coleoptera. Academic Press, London.

Csiki, E. 1927-1933. Coleopterorum Catalogus. Carabidae, parts 91, 92, 97, 98, 104, 112, 115, 121, 124, 126-W. Junk, Berlin.

Csiki, E. 1927b. Carabidae: Carabinae. *In:* Coleopterorum Catalogus (Ed.) S. Schenkling (Parts 91 and 92). Junk Publ., 621 p.

Curcic, S. and Stankovic, M. 2011. The Ground beetles (Coleoptera: Carabidae) of the Zasavica Special Nature Reserve (Serbia). *Acta Entomologica Serbica*, **16(1/2):** 61–79.

Dambach, Charles Arthur. 1948. A study of the ecology and economic value of crop field borders. Graduate School Studies. *Biological Science Series*, **2:** 1–205.

Darlington, P.J. 1943. Carabidae of mountains and islands: Data on the evolution of isolated faunas, and atrophy of wings. *Ecological Monograph*, **13:** 37–61.

Darlington, P.J. 1961a. Australian carabid beetles. V. Transition of wet forest faunas from New Guinea-Tasmania. *Psyche*, **68:** 1–24.

Darlington, P.J. 1970. Insects of Micronesia (Coleoptera: Carabidae including Cicindellidae). **15(1):** 1–49.

Davies, L.1987. Long adult life, low reproduction and competition in two sub-Antractic carabid beetle. *Ecological Entomology*, **12:** 149–162.

Dawson, N. 1965. A comparative study of the ecology of eight species of fenland Carabidae (Coleoptera). *Journal of Animal Ecology*, **34:** 299–314.

Day, K.R. 1987. The species and community characteristics of ground beetles (Coleoptera: Carabidae) in some Northern Ireland nature reserves. *Proceedings of the Royal Irish Academy*, **87B:** 65–82.

Dearborn, R.G., Nelson, R.E., Donahue, C., Bell, R.T. and Webster, R.P. 2014. The Ground Beetle (Coleoptera: Carabidae) Fauna of Maine, USA. *The Coleopterists Bulletin*, **68:** 441–599.

Debach, P. and Rosen, D. 1991. Biological Control by Natural Enemies (2nd edition). Cambridge University Press. 408 p.

Deleurance, S. and Deleurance, E.P. 1964. Reproduction et cycle evolutif larvaire des *Aphenops* (*A. cerberus* Dieck, *A. crypticola* Linder) insectes

Coleopteres cavernicoles. *Comptes rendus de l'Académie des Sciences Paris*, **258:** 4369–4370.

Deng, D.A. and Li, B.Q. 1981. Collecting ground beetles (Carabidae) in baited pitfall traps. *Insect Knowledge*, **18:** 205–207. (In Chinese)

Desender, K. 1983. Ecological data on *Clivina fossor* (Coleoptera, Carabidae) from a pasture ecosystem. 1. Adult and larval abundance, seasonal and diurnal activity. *Pedobiologia*, **25:** 157–167.

Desender, K., Maelfait, J.P., D'Hulster, M. and Vanhercke, L. 1981. Ecological and faunal studies on Coleoptera in agricultural land. I. Seasonal occurrence of Carabidae in the grassy edge of a pasture. *Pedobiologia*, **22:** 379–384.

van Dijk, T.S. 1972. The significance of the diversity in age composition of *Calathus melanocephalus* L. (Coleoptera, Carabidae) in space and time at Schiermonnikoog. *Oecologia*, **10:** 111–136.

van Dijk, T.S. 1979a. On the relationship between reproduction, age and survival in two carabid beetles *Calathus melanocephalus* L. and *Pterostichus coerulescens* L. (Coleopteran, Carabidae). *Oecologia*, **40:** 63–80.

van Dijk, T.S. 1979b. Reproduction of young and old females in two carabid beetles and the relationship between the number of eggs in the ovaries and the number of eggs laid. *Miscllaneous Papers Landbouw Wageningen*, **18:** 167–183.

van Dijk, T.S. 1983. The influence of food and temperature on the amount of reproduction in carabid beetles. *In:* P. Brandmayr, P.J. den Boer and E. Weber (eds). Ecology of Carabids: The Synthesis of Field Study and Laboratory Experiment. Centre Agric Publ. Doc. Wageningen.

van Dijk, T.S. 1986. How to estimate the level of food availability in field population of carabid beetles. *In:* P.J. den Boer, M.L. Luff, D. Mossokowski and F. Weber (eds). Carabid Beetles: Their Adaptations and Dynamics. G. Fischer. Stuttgart.

van Dijk, T.S. 1994. On the relationship between food, reproduction and survival of two carabid beetles: *Calathus melanocephalus* and *Pterostichus versicolor. Ecological Entomology*, **19:** 263–270.

Doane, J.F. 1981. Seasonal captures and diversity of ground beetles (Coleoptera: Carabidae) in a wheat field and its grassy border in central Saskatchewan. *Quaestiones Entomologicae*, **17:** 211–233.

Dowdy, W.W. 1947. An ecological study of the Arthropoda of an oak-hickory forest with reference to stratification. *Ecology*, **28:** 418–439.

Dreisig, H. 1981. The rate of predation and its temperature dependence in a tiger beetle, *Cicindela hybrid. Oikos*, **36:** 196–202.

Duff, A.G. 1992. Some old records of rare Carabidae (Col.) from western Britain. *Entomologist's Monthly Magazine*, **128:** 37.

Duff, A.G. 1993. Beetles of Somerset: Their Status and Distribution. Taunton: Somerset Archaeological and Natural History Society. Tounten, 269 p.

Dunning, R.A., Baker, A.N. and Windley, R.F. 1975. Carabid in sugarbeat and their possible role as aphid predators. *Annals of Applied Biology*, **80:** 125–128.

Edmonds, T.H. 1934. A new species of *Tachys* (Coleoptera, Carabidae) from the New Forest, new to science (*Tachys piceus* n. sp.). *Entomologists' Monthly Magazine*, **70:** 7–9.

Edwards, C.A., Sunderland, K.D. and George, K.S. 1979. Studies on polyphagous predators of cereal aphids. *Journal of Applied Ecology*, **16:** 811–823.

Ellsbury, M.M., Powell, J.E., Forcella, F., Woodson, W.D., Clay, S.A. and Riedell, W.E. 1998. Diversity and dominant species of ground beetle assemblages (Coleoptera: Carabidae) in crop rotation and chemical input systems for the northern Great Plains. *Annals of the Entomological Society of America*, **91:** 619–625.

Else, G.R. 1993. The distribution and habitat requirements of the tiger beetle *Cicindela germanica* Linnaeus (Coleoptera Carabidae) in southern Britain. *British Journal of Entomology and Natural History*, **6:** 17–21.

Ernoult, A., Vialatte, A., Butet, A., Michel, N., Rantier, Y., Jambon, O. and Burel, F. 2013. Grassy strips in their landscape context, their role as new habitat for biodiversity. *Agricultural Ecosystem and Environment*, **166:** 15–27.

Erwin, T.L. 1965. A revision of Brachinus of North America. *The Coleopterists' Bulletin*, **19(1):** 1–19.

Erwin, T.L. 1970. A reclassification of bombardier beetle and a taxonomic revision of the North and Middle American species (Carabidae: Brachinida). *Quaestiones Entomologicae*, **6:** 4–215.

Erwin, T.L. 1971. Notes and corrections to a reclassification of bombardier beetle (Carabidae, Brachinida). *Quaestiones Entomologicae*, **7:** 281.

Erwin, T.L. 1974. Studies of the subtribe *Tachyina* (Coleoptera: Carabidae: Bembidiini) supplement A; Lectotype designations for new world species, two new genera, and notes on generic concept. *Proceedings of the Entomological Society of Washington*, **76(2):** 123–154.

Erwin, T.L. 1979a. Thoughts on the evolutionary history of ground beetles: Hypotheses generated from comparative faunal analysis of lowland forest sites in temperate and tropical regions. *In:* T.L. Erwin, G.E. Ball, D.L. Whitehead and A.L. Halpen (eds). Carabid Beetles: Their Evolution, Natural History and Classification. Junk, The Hague, pp. 539–592.

Erwin T.L. 1979b. A review of the natural history and evolution of ectoparasitoid relationship in carabid beetles. *In:* T.L. Erwin, G.E. Ball, D.L. Whitehead and A.L. Halpen (eds). Carabid Beetles: Their Evolution, Natural History and Classification,The Hague, pp. 479–484.

Erwin, T.L. 1985. The taxon pulse: A general pattern of lineage radiation and extinction among carabid beetles. *In:* G.E. Ball (ed.). Taxonomy, Phylogeny and Zoogeography of Beetles and Ants. Junk, The Hague pp. 437–472.

Erwin, T.L. 2008. A Review of *Gallerucidia chaudoir* (Carabidae: Lebiini: Gallerucidiina): Their occurrences, ways of life, and another genus record for Costa Rica. *Annals of Carnegie Museum*, **77(1):** 135–146.

Erwin, T.L. and Adis, J. 1982. Amazonian inundation forests: Their role as short-term refuges and generators of species richness and taxon pulses. *In:* G. Prance (ed.). Biological Diversification in the Tropics. Columbia University Press, New York.

Erwin, T.L. and Halpern, A.L. 1978. Max Banninger: His collection and publications (Coleoptera, Carabidae). *The Coleopterists Bulletin*, **32(4):** 357–386.

Erwin, T.L. and Stork, N.E. 1985. The Hiletini, an ancient and enigmatic tribe of Carabidae with a pantropical distribution (Coleoptera). *Systematic Entomology*, **10:** 405–451.

Esau, K.L. 1968. Carabidae (Coleoptera) and Other Arthropods Collected in Pitfall Traps in IOWA Cornfields, Fencerows and Prairies. A Dissertation submitted to the Graduate Faculty in Partial Fulfilment of the Requirements for the Degree of Doctor of Philosophy in IOWA State University of Science and Tecnology Ames, IOWA. 169 p.

Esau, K.L. and Peters, D.C. 1975. Carabidae collected in pitfall traps in Iowa cornfields, fencerows, and prairies. *Environmental Entomology*, **4:** 509–513.

Eubanks, M.D. and Denno, R.F. 2000. Host plants mediate omnivore-herbivore interactions and influence prey suppressions. *Ecology*, **81(4):** 936–947.

Evans, M.E.G. 1977. Locomotion in the Coleoptera Adephaga, especially Carabidae. *Journal of Zoology*, **181:** 189–226.

Evans, M.E.G. 1986. Carabid locomotor habits and adaptations. *In:* P.J. den Boe, M.L. Luff, D. Mossakowski and F. Weber (eds). Carabid Beetles: Their Adaptation and Dynamics. Fischer Verlag, Stuttgart/New York.

Evans, W.G. 1988. Chemically mediated habitat recognition in shore insects (Coleoptera: Carabidae; Hemiptera: Saldidae). *Journal of Chemical Ecology*, **14:** 1441–1454.

Everly, Ray Thomas. 1938. Spider and insects found associated with sweet corn with notes on the food and habit of some species. I. Arachnida and Coleoptera. *Ohio Journal of* Science, **38:** 136–148.

Eyre, M.D., Luff, M.L. and Leifert, C. 2013. Crop field boundary productivity and disturbances on ground beetle (Coleoptera, Carabidae) in the agroecosystem. *Agriculture Ecosystem and Environment*, **165:** 60–67.

Facchini, S. and Sciaky, R. 2004. Two new species of Bradybaenus Dejean, 1829 from Africa (Coleoptera: Carabidae: Harpalinae). *Koleopterologische Rundschau*, **74:** 11–24.

Fazekas, J.P., Kadar, F., Sarospataki, M. and Lovei, G.L. 1997. Seasonal activity, age structure and egg production of the ground beetle

Anisodactylus signatus (Coleoptera, Carabidae) in Hungary. *European Journal of Entomology*, **94:** 485–494.

Fedorenko, D.N. 2014. New or little –known Pericaline and Cymindidina (Coleoptera: Carabidae from the Oriental region. *Russian Entomological Journal*, **23(4):** 305–315.

Fletcher, T.B. 1919. Second hundred notes. *Agricultural Research Institute. Pusa Bulletin*, **89:** 31–35.

Forbes, S.A. 1880. Notes on insectivorous Coleoptera. *Illinois State Laboratory Natural History Bulletin*, **1(3):** 153–160.

Forbes, S.A. 1883. The food relations of the Carabidae and the Coccinellidae. *Bulletin of the Illinois State Laboratory of Natural History*, **1(6):** 33–64.

Forrest, T.G. 1985 Reproductive behavior. *In:* I.J. Walker (ed.). Mole crickets in Florida. *Florida Agricultural Experiment Station Bulletin*, **846:** 10–15.

Fowler, W.W. 1887. The Coleoptera of the British Islands. Vol. 1. London: Reeve.

Fowles, A.P. 1989. The Coleoptera of shingle banks on the River Ystwyth, Dyfed. *Entomologist's Record*, **101:** 209–221.

Fowles, A.P. and Boyce, D.C. 1992. Rare and notable beetles from Cardiganshire (VC46) new to Wales. *Coleopterist*, **1(1):** 7–8.

Frank, J.H. 1971. Carabidae (Coleoptera) of an arable field in central Alberta. *Quaestiones Entomologicae*, **7:** 237–252.

Frank, S.D. 2007. Consequences of Omnivore and Alternative Food Resources on the Strength of Trophic Cascades. Dissertation of Ph.D. thesis, University of Maryland, USA.

Frank, J.H., Erwin, T.L. and Hemenway, R.C. 2009. Economically beneficial ground beetles. The specialized predators *Pheropsophus aequinoctialis* (L.) and *Stenaptinus jessoensis* (Morawitz): Their laboratory behavior and descriptons of immature stages (Coleoptera, Carabidae, Brachininae) *Zookeys*, **14:** 1–36.

Freitag, R. 1969. Revision of the genus *Evarthrus* LeConte. *Quaestiones Entomologicae*, **5:** 89–212.

Freitag, R. 1979. Carabid beetles and pollution. *In:* T.L. Erwin, G.E. Ball and D.R. Whitehead (eds). Carabid Beetles: Their Evolution, Natural History, and Classification. Dr. W. Junk, The Hague, The Netherlands.

French, B.W. and Elliott, N.C. 1999a. Temporal and spatial distribution of ground beetle assemblages in grasslands and adjacent wheat fields. *Pedobiologia*, **43:** 73–84.

French, B.W. and Elliott, N.C. 1999b. Spatial and temporal distribution of ground beetle assemblages in riparian strips and adjacent wheat fields. *Environmental Entomology*, **28:** 597–607.

French, B.W., Elliott, N.C. and Berberet, R.C. 1998. Reverting conservation reserve program lands to wheat and livestock production: Effects on ground beetle (Coleoptera: Carabidae) assemblages. *Environmental Entomology*, **27:** 1323–1335.

French, B.W., Elliott, N.C., Berberet, R.C. and Burd, J.D. 2001. Effects of riparian and grassland habitats on ground beetle (Coleoptera: Carabidae) assemblages in adjacent wheat fields. *Environmental Entomology*, **30:** 225–234.

French, B.W., Chandler, L.D., Ellsbury, M.M., Fuller, B.W. and West, M. 2004. Ground Beetle (Coleoptera: Carabidae) Assemblages in a Transgenic Corn-Soybean Cropping system. *Environmental Entomology*, **33(3):** 554–563.

Freude, H., Harde, K.W. and Loba, G.A. 1976. *Die Kaefer Mitteleuroras. Band 2, Adephaga 1*. Goecke and Evers Krefeld.

Fuller, B.W. 1988. Predation by *Calleida decora* (F.) (Coleoptera: Carabidae) on velvetbean caterpillar (Lepidoptera: Noctuidae) in soybean. *Journal of Economic Entomology*, **81(1):** 127–129.

Gamboa, S. and Ortuño, Vicente M. 2015. A new fossil species of the genus Coptodera Dejean, 1825 (Coleoptera: Carabidae: Lebiinae) from Baltic amber. *Zootaxa*, **3981(4):** 592–596.

Ganglbauer, L. 1905. Coleoptera. *In:* A. Penther und E. Zrderbauer (Edr.), Ergebnisse einer naturwissenschaftlichen Reise zum Erddschias-Dagh (Kleinasien). *Annalen des K.K. Naturhistorischen Hofmuseums*, Wien, **20(2-3):** 246–290.

George, E.B. 1996. Vignettes of the history of neotropical carabidology (PDF). *Annales Zoologici Fennici*, **33:** 5–16.

Gergely, G. and Lovei, G.L. 1987. Phenology and reproduction of the ground beetle *Dolichus halensis* in maize fields: A preliminary report. *Acta Phytopathologica et Entomologica Hungarica*, **22:** 357–361.

Ghahari, H., Kesdek, M., Samin, N., Ostvan, H., Havaskary, M. and Imani, S. 2009. Ground beetles (Coleoptera: Carabidae) of Iranian cotton fields and surrounding grasslands. *Munis Entomology & Zoology*, **4(2):** 436–450.

Ghahari, H. and Kesdek, M. 2012. A study on the species diversity of ground beetles (Coleoptera: Carabidae) from Khoasan province, Eastern Iran. *Entomofauna*, **33:** 1–8.

Gidaspow, T. 1959. North American caterpillar hunters of the genera *Calosoma* and *Callisthenes* (Coleoptera: Carabidae). *American Museum of Natural History Bulletin*, **116(3):** 229–343.

Gilbert, O. 1946. The life history of *Nebria degenerata* Schaufuss and *Nebria brevicollis* Fab. (Coleoptera, Carabidae). *Journal of the Society for British Entomology*, **6:** 11–14.

Gregoire-Wibo, C. 1983. Incidence écologique de traitements phytosanitaires en culture de betteraves sucrière. II. Acariens, polydesmes, staphylins, cryptophagides et carabides. *Pedobiologia*, **25:** 93–108.

Griffiths, E., Wratten, S.D. and Vickerman, G.P. 1985. Foraging by the carabid *Agonum dorsale* in the field. *Ecological Entomology*, **10:** 181–189.

Grum, L. 1973. Egg production of some Carabidae species. *Bulletin de l' Académie Polonaise des Sciences. Série des Sciences Biology*, **21:** 261–268.

Grum, L. 1984. Carabid fecundity as affected by extrinsic and intrinsic factors. *Oecologia*, **65:** 114–121.

Guéorguiev, V.B. and Guéorguiev, B.V. 1995. Catalogue of the Ground-Beetles of Bulgaria (Coleoptera: Carabidae). Pensoft Publishers Sofia, Bulgaria. 279 p.

Guéorguiev, V.B. 2014. *Eustra petrovi* sp. Nov. – First record of a troglobitic Ozaenini from China (Coleoptera: Carabidae: Paussinae). *Journal of Insect Biodiversity*, **2(10):** 1–9.

Gutiérrez, D., Menéndez, R. and Méndez, M. 2004. Habitat-based conservation priorities for carabid beetles within the Picos de Europa National Park, northern Spain. *Biological Conservation*, **115:** 379–393.

Habu, A. and Sadanaga, K. 1961. Illustrations for identifications of larvae of the Carabidae found in cultivated fields and paddy fields (I). *Bulletin of the National Institute of Agricultural Science Series C* (Plant Pathology Entomology), **13:** 207–248.

Habu, A. and Sadanaga, K. 1965. Illustrations for identification of larvae of the Carabidae found in cultivated fields and paddy fields (III). *Bulletin of the National Institute of Agricultural Science* (Japan) *Series C*, **19:** 81–216 [In Japanese see 172–177].

Habu, A. and Sadanaga, K. 1969. Illustrations for identification of larvae of the Carabidae found in cultivated fields and paddy fields (Suppl. I). *Bulletin of the National Institute of Agricultural Sciences* (Japan) *Series C*, **23:** 113–143 [in Japanese see 133–135].

Häckel, M. and Kirschenhofer, E. 2014b. A Contribution to knowledge of the subfamily Panagaeinae Hope, 1838 from Asia. Part 2. East Palearctic and Oriental Species of the genus *Craspedophorus* Hope, 1838, and the genus *Tinoderus chaudoir*, 1879 (Coleoptera: Carabidae) // Studies and Reports. *Taxonomical series*, **10(2):** 275–392. Jedlička A. 1965. Monographie des tribus Panagaeini aus Ostasien

Haddad, N.M., Haarstad, J. and Tilman, D. 2000. The effect of long-term nitrogen loading on grassland insect communities. *Oecologia*, **124:** 73–84.

Haeck, J. 1971. The immigration and settlement of Caribidae in the new Ijsselmeerpolders. *In:* P.J. den Boer (ed.). Dispersal and dispersal power of carabid beetles. *Miscllaneous Paper Landbouwhogesch Wageningen*, **8:** 33–52.

Hance, T. 1987. Predation impact of carabits of different population densities on the development of *Aphis fabae* in sugar beet. *Pedobiologia*, **30:** 251–262.

Hance, T. 2002. Impact of cultivation and crop husbandry practices. *In:* J.M. Holland (ed.). The Agroecology of Carabid Beetle. pp. 231-239. Andover: Intercept Ltd.

Henderson, M. 1991. *Polistichus connexus* (Fourcroy) (Coleoptera: Carabidae) on Wimbledon Common. *British Journal of Entomology and Natural History*, **4:** 8-8.

Hansen, M. 1996. Catalogue of the Coleoptera of Denmark. *Entomologiske Meddelelser*, **64:** 1–231.

Hatch, M.H. 1953. Studies on the Coleoptera of the Pacific Northwest. IV. Carabidae, Dytiscidae, Gyrinidae. *Brooklyn Entomological Society Bulletin*, **46:** 113–122.

Hayward, R. 1908. Studies in *Amara. American Entomological Society Transactions*, **34:** 13–65.

Heer, O. 1847. Die Insektenfauna der Tertiargebilde von Oeningen und von Radoboj in Croatien. *Erster Theil*: Kafer 1–229.

Heer, O. 1861. Recherches sur le climat et la vegetation du pays tertiaire. Paris, Wurster and Co.

Heinrichs, E.A. 1994. Biology and Management of Rice Insect. IRRI, Wiley Eastern Limited. 779 p.

Heinrichs, E.A. and Barrion, A.T. 2004. Rice Feeding Insect and Their Natural Enemies in West Africa – Biology, Ecology, Identification. IRRI, Philippines. 242 p.

Hendrickson, G.O. 1930. Studies on the insect fauna of Iowa prairies. *Iowa State College Journal of Science*, **4:** 49–179.

Hengeveld, R. 1980b. Polyphagy, Oligophagy and food specialization in ground beetles (Coleoptera: Carabidae). *Netherlands Journal of Zoology*, **30:** 564–584.

Herne, D.H.C. 1963. Carabids collected in a DDT-sprayed peach orchard in Ontario (Coleoptera: Carabidae). *Canadian Entomologist*, **95:** 357–362.

Hessen, H.J.L. 1980. Egg production of *Pterostichus oblongpunctatus* (Fab.) (Col. Carabidae) and *Philonthus decorus* (Gravenhorst) (Col. Staphylinidae). *Netherlands Journal of Zoology*, **30:** 35–53.

Hessen, H.J.L. 1981. Egg mortality in *P. oblogopunctatus* (Coleoptea, Carabidae). *Oecologia*, **50:** 233–235.

Heydemann, B. and Meyer, H. 1983. Auswirkungen der Intensivkultur auf die Fauna in grarbiotopen. *Schriftenreihe Deutscher Rat fiir Landespflege*, **42:** 174–191.

Heyler, N., Cattlin, N. and Brown, K. 2009. Biological Control in Plant Protection. CRC Press 133–136 (ebook_promo.png).

Hieke, F., Kavanaugh, D.H. and Liang, H. 2012. A new species of *Amara* (Coleoptera, Carabidae, zabrini) from Sichuan Province, China with additional records for other *Amara* species from the region. *Zookeys*, **254:** 47–65. doi:10.3897/zookeys.254.4223. (Accessed on Aug. 7, 2018)

Hill, D.S. 1993. Agricultural Insect pests of the Tropics and Their Control. Cambridge University Press. 746 p.

Hinton, H.E. 1945. A Monograph of the Beetles Associated with Stored Products. Vol. 1. London: British Museum.

Hogan, J.E. 2012. Taxonomy, Systematic and Biogeography of the Scaritinae (Insecta, Coleoptera, Carabidae) Ph.D. Thesis. Oxford Brookes University.

Hokkanen, H. and Holopainen, J.K. 1986. Carabid species and activity densities in biologically conventionally managed cabbage field. *Journal of Applied Entomology*, **102:** 353–363.

Holliday, N.J. and Hagley, E.A.C. 1984. The effect of sod type on the occurrence of ground beetles (Coleoptera: Carabidae) in a pest management apple orchard. *Canadian Entomologist*, **116:** 165–171.

Honek, A. and Jorosik, V. 2000. The role of crop density, seed and aphid presence in diversification of field communities of Carabidae (Coleoptera). *European Journal of Entomology*, **97:** 517–525.

Honek, A., Martinkova, Z. and Jarosik, V., 2003. Ground beetles (Carabidae) as seed predators. *European Journal of Entomology*, **100:** 531–544.

Horne, P.A. 1990. Parental care in *Notonomus* Chaudoir (Coleoptera: Carabidae: Pterostichinae). *Australian Entomological Magazine*, **17:** 65–69.

Horvatovich, S. and Szarukan, I. 1981. Contribution a la biologie et morphologie des especes hongroises du genre *Anisodactylus* Dejean (Coleoptera: Carabidae). *Journal of Muzeum Evkönyve*, **26:** 13–27.

http://animal diversity.org/ (Accessed on January 12, 2017)

http: // bugguide.net/node/view/10161 (Accessed on June 12, 2017)

http://carabidae.org/taxa/albipes-fabricius (Accessed on February 14, 2017)

http://carabidae.org/taxa/bilunatus-burgeon (Accessed on July 13, 2016)

http://carabidae.org/taxa/binotata-dejean (Accessed on March 13, 2017)

http://carabidae.org/taxa/bisulcatus-nicolai (Accessed on December 10, 2017)

http://carabidae. org/taxa/ calycophora-schmidtgoebel (Accessed on December 10, 2017)

http://carabidae.org/taxa/fuscomaculatus-motschulsky (Accessed on April 5, 2017)

http://carabidae.org/taxa/javanus-klug- (Accessed on August 15, 2017)

http://carabidae.org/taxa/macrocheilus (Accessed on January 5, 2017)

http://carabidae.org/taxa/nana-gyllenhal-1810 (Accessed on January 12, 2017)

http://carabidae.org/taxa/polita-faldermann- (Accessed on January 18, 2016)

http //www.carabidae.ru/ (Accessed on January 10, 2016)

http://www.center for systematic entomology.org/. (Accessed on August 8, 2016)

http://creative-commons.org. (Accessed on January 20, 2017)

http://data1.among leaf litter,rottinglog (Accessed on July 16, 2017)

http://ecoregister.org/) (Accessed on March 10, 2017)

https://en.wikipedia.org/wiki/Aepus (Accessed on March 8, 2017)

https://en.wikipedia.org/wiki/Laemostenus (Accessed on November 9, 2016)

https://en.wikipedia.org/wiki/Perileptus_areolatus (Accessed on March 12, 2017)

https//en.wikipedia.org/wiki/scarities (Accessed on September 10, 2017)

http: //ent.psu.edu/extension/factsheets/groundbeetle (Accessed on February 10, 2017)

http://eol.org (Accessed on February 18, 2017)
http//extension.umaine.edu (Accessed on December 9, 2016)
http://galerie.insect.org (Accessed on January 1, 2016)
http://nature.berkeley.edu/ (Accessed on February 7, 2017)
http://taibif.tw/en/catalogue (Accessed on July 10, 2016)
http://www.habitas.org.uk (Accessed on August 5, 2016)
http://www.habitas.org.uk/groundbeetles (Accessed on November 8, 2017)
http//www.minibeastwildlife.com.all/research/tiger-beetle (Accessed on February 10, 2017)
https//www. planetnatural.com. (Accessed on February 7, 2016).
http//www.sibnef1.eu/gb/coleopteran/carabidae/img100/eco. (Accessed on December 7, 2016)
http//www.wikivisually.com/wiki/Abacidus (Accessed on July 10, 2017)
House, G.J. and All, J.N. 1981. Carabid beetles in soybean agroecosystems. *Environmental Entomology*, **10:** 194–196.
Houston, W.W.K. 1981. The life cycles and age of *Carabus glabratus* Paykul and *C. probletnaticus* Herbst (Col.: Carabidae) on moorland in northern England. *Ecological Entomology*, **6:** 263–271.
Houston, W.W.K. and Luff, M.L. 1983. The identification and distribution of the three species of *Patrobus* (Coleoptera: Carabidae) found in Britain. *Entomologist's Gazette*, **34:** 283–288.
van Huizen, T.H.P. 1977. The significance of flight activity in the life cycle of *Amara plebeja* Gyll. (Coleoptera, Carabidae). *Oecologia* (Berlin), **29:** 27–41.
van Huizen, T.H.P. 1979. Individual and environmental factors determining flight in carabid beetles. *In:* P.J. den Boer, H.U. Thiele and F. Weber (eds). On the evolution of behaviour in carabid beetles. *Miscellaneous Paper. Agricultural University of Wageningen*, **18:** 199–211.
van Huizen, T.H.P. 1990. Gone with the wind: Flight activity of carabid beetles in relation to wind direction and to the reproductive state of females flight. *In:* N.E. Stork (ed.). The Role of Ground Beetles in Ecological and Environmental Studies. Intercept, Andover.
Humbert, J.Y., Ghasoul, J. and Walter, T. 2008. Meadow harvesting techniques and their impacts on field fauna. *Agriculture, Ecosystems and Environment,* **130:** 1–8.
Hurka, K. 1996. Carabidae of the Czech and Slovak Republics. Kabourek Publishing, Zlin, Czech Republic. 565 p.
Hyman, P.S. and Parsons, M. 1992. A review of the scarce and threatened Coleoptera of Great Britain. Part 1. Peterborough: Joint Nature Conservation Committee.
Ito, N. and Jaeger, B. 2000. A description of new species of the Bradycelline subgenus Tachycellus from India and Nepal with notes on the *Bradycellus anchomenoides* Group (Coleoptera: Carabidae: Harpalini). *Linzer Biologische Beiträge*, **32/2:** 1215–1223.

Jaques, H.E. and Redlinger, L. (1946). A preliminary list of the Carabidae known to occur in Iowa. *Proceedings of the Iowa Academy of Science*, **52:** 293–298 [DP: July 1946].

Jensen, T.S., Drying, L., Kristensen, B., Nielsen, B.O. and Rasmussen, E.R. 1989. Spring dispersal and summer habitat distribution of *Agonum dorsale* (Coleoptera: Carabidae). *Pedobiologia*, **33:** 155–165.

Jobe, J.B. 1990. *Leistus rofomarginatus* (Duftschmid) (Col., Carabidae) flying in North Yorkshire. *Entomologist's Monthly Magazine*, **126:** 200.

Jorgensen, H.B. and Lovei, G.L. 1999. Tri-trophic effect on predator feeding: Consumption by the carabid *Harpalus affinis* of *Heliothis armigera* caterpillars fed on proteinase inhibitor-containing diet. *Entomologia Experimentalis et Applicata*, **93:** 113–116.

Jørgensen, H.B. and Toft, S. 1997. Food preference, diet dependent fecundity and larval development in *Harpalus rufipes* (Coleoptera: Carabidae). *Pedobiologia*, **41:** 307–315.

Joy, N.H. 1932. A Practical Handbook of British Beetles. London: Witherby.

Juliano, S.A. 1985. Habitat associations, resources, and predators of an assemblage of Brachinus (Coleoptera; Carabidae) from southeastern Arizona. *Canadian Journal of Zoology*, **62:** 1683–1691.

Kasandrova, L.I. and Sharova, I.H. 1971. Development of *Amara ingénua, Anisodactylussignatus* and *Harpalus distinguendus* (Coleoptera, Carabidae). *Zoologicheskii Zhurnal*, **50:** 215–221 (in Russian).

Kataev, B.M. 2012. Species of the genus Dioryche similar to *D. cuprina* (Dejean, 1929) comb. nov. (Coleoptera: Carabidae: Harpalini). *Zoosystematica Rossica*, **21(1):** 112–130.

Kennedy, P. 1994. The distribution and movement of ground beetles in relation to set-aside arable land. *In:* K. Desender, M. Dufrene, M.L. Cuff and J.-P. Maelfait (eds). Carabid Beetles: Ecology and Evolution. Kluwer Acdemic, Dordrecht.

Kesdek, M. 2012. A contribution to the knowledge of the Carabidae (Coleoptera) fauna of Turkey. *Acta Biologica Universitatis Daugavpiliensi*, **12(1):** 55–56.

Kesdek, M. 2015. New data on distribution of ground beetles of the tribe Harpalini in the Palaearctic Oriental Region and in Australia (Coleopteran, Carabidae, Harpalini). *Entomological Review*, **95(4):** 536–543.

Kesdek, M. and Yildirim, E. 2004. Contribution to the knowledge of Carabidae fauna of Turkey. Part 2: Platynini (Coleoptera, Carabidae). *Linzer Biologische Beiträge*, **36(1):** 527–533.

Kesdek, M. and Yildirim, E. 2007a. Contribution to the knowledge of Carabidae fauna of Turkey. Part 3: Bembidiini (Coleoptera, Carabidae, Bembidiinae), *Entomofauna*, **28:** 117–124.

Kesdek, M. and Yildirim, E. 2007b. Contribution to the knowledge of Carabidae fauna of Turkey. Part 4: Dryptini, Lebiini and Zuphiini (Coleoptera, Carabidae, Lebiinae), *Entomofauna*, **28:** 277–284.

Kesdek, M. and Yildirim, E. 2010a. Contribution to the knowledge of Carabidae fauna of Turkey. Part 6: Notiophilini (Notiophilinae) and Platynini (Pterostichinae) (Coleoptera, Carabidae). *Entomofauna*, **31:** 5–16.

Kesdek, M. and Yildirim, E. 2010b. Contribution to the knowledge of Carabidae fauna of Turkey. Part 8: Amarini (Coleopteran, Carabidae, Pterostichinae), *Entomofauna*, **31:** 17–24.

Key, R.S. 1996. *Stenolophus skrimshiranus* Stephens and *Badister unipustulatus* Bonelli (Carabidae) in South Lincolnshire. *Coleopterist*, **5:** 22.

King, P.E. and Stabins, V. 1971. Aspects of the biology of a strand-living beetle, *Ewynebria complanata* (L.). *Journal of Natural History*, **5:** 17–28.

Kirk, V. M. 1971. Ground beetles in cropland in South Dakota. *Annals of the Entomological Society of America*, **64:** 238–241.

Kirschenhofer, E. 2010. New and little-known species of Carabidae from the Middle East and Southeast Asia (Coleoptera, Carabidae: Lebiini, Brachinini). *Annales Historico-Naturales Musei Nationalis Hungarici*, **102:** 25–64.

Kobayashi, M., Kudagamage, C. and Nugaliyadde, L. 1995. Distribution of larvae of *Ophioneaindica* Thunberg (Carabidae), a predator of the Rice Gall Midge, *Orseolia oryzae* (Wood-Mason) in paddy fields of Sri Lanka. *Japan Agricultural Research Quarterly*, **29:** 89–93.

Kocatepe, N. and Mergen, O. 2004. The faunistical studies on the family Carabidae (Coleoptera) species in Ankara province. *Türkiye Entomoloji Dergisi*, **28(4):** 295–309.

Kreckwitz, H. 1980. Experiments in the breeding biology and the seasonal behavior of the carabid beetle *Agonum dorsale* Pont. in temperature and moisture gradients. *Zoologische Jahrbücher. Abteilung für Systematik*, **107:** 183–234.

Kromp, B. 1989. Carabid beetle communities (Carabidae, Coleoptera) in biologically and conventionally farmed agroecosystems. *Agriculture, Ecosystems and Environment*, **27:** 241–251.

Kromp, B. 1990. Carabid beetles (Coleoptera, Carabidae) as bioindicators in biological and conventional farming in Austrian potato fields. *Biology and Fertility of Soils*, **9:** 182–187.

Kromp, B. 1999. Carabid beetles in sustainable agriculture: A review of pest control efficacy, cultivation impacts and enhancement. *Agriculture, Ecosystems and Environment*, **74:** 187–228.

Kryzhanovskij, O.L., Belousov, I.A., Kabak, I.I., Kataev, B.M., Makarov, K.V. and Shilenkov, V.G. 1995. A Checklist of the Ground Beetles of Russia and Adjacent Lands (Insecta, Coleoptera, Carabidae). Pensoft Publish., Sofia, Bulgaria. 271 p.

Kumar, P. 1997. Distribution of Carabid beetles (Coleoptera: Carabidae) in Karnataka with notes on their habitats and feeding habitats. *Karnataka Journal of Agricultural Sciences*, **10(4):** 991–998.

Kumar, P. and Rajagopal, D. 1990. Carabid beetle, *Omphra pilosa* Klug. (Coleoptera: Carabidae) a potential predator on termites. *Journal of Biological Control*, **4(2):** 105–108.

Kumar Prasad and Rajagopal, D. 1997. Carabid Fauna (Coleoptera: Carabidae: Harpalini) of Karnataka and their ecology. *Karnataka Journal of Agricultural Sciences*, **10(2):** 322–325.

Kushwaha, R.K. and Hegde, V.D. 2012. Records of the Zoological Survey of India, **112(Part-2):** 119–120.

Kushwaha, R.K., Hegde, V.D. and Chandra, K. 2015. Ground Beetles (Coleoptera: Carabidae) from Chhattisgarh, India. *Zoos' Print*, Volume XXX, Number 1.

Landis, D.A., Wratten, S.D. and Gurr, G.M. 2000. Habitat management to conserve natural enemies of arthropod pests in agriculture. *Annual Review of Entomology*, **45:** 175–201.

Lang, A. 2003. Intraguild interference and biocontrol effects of generalist predators in a winter wheat field. *Oecologia*, **134:** 144–153.

Lapshin, L.V. 1971. Seasonal activity of dominant species of ground beetle (Carabidae) in the forest-steppe zone near Ohrenburg. *Zoologicheskii Zhurna*, **50:** 825–833 (in Russian).

Larochelle, A. 1975a. A list of mammals as predators of Carabidae. *Carabologia*, **3:** 95–98.

Larochelle, A. 1975b. A list of amphibians and reptiles as predators of Carabidae.*Carabologia*, **3:** 99–103.

Larochelle, A. 1980. A list of birds of Europe and Asia as predators of carabid beetles including Cicindelini (Coleoptera: Carabidae). *Cordulia*, **6:** 1–19.

Larochelle, A. 1990. The food of Carabid Beetles (Coleoptera: Carabidae, including Cicindelinae). *Fabreries*, Supplement **5:** 1–132

Larochelle, A. and Larivière, M.C. 2003. A Natural History of the Ground-Beetles (Coleoptera: Carabidae) of America, North of Mexico. Pensoft Sofia, Bulgaria.

Larochelle, A. and Larivière, M.C. 2005. Harpalini (Insecta: Coleoptera: Carabidae: Harpalinae). *Fauna of New Zealand*, **53:** 160.

Larochelle, A. and Larivière, M.C. 2013. Carabidae (Insecta: Coleoptera): Synopsis of species Cicindelinae to Trechinae (in part). *Fauna of New Zealand*, **69:** 193.

Larsen, J.K., Work, T.T. and Purrington, F.F. 2003. Habitat use patterns by ground beetles (Coleoptera: Carabidae) of North Eastern Lowa. *Pedobiologia*, **47:** 288–299.

Laub, C.A. and Luna, J.M. 1992. Winter cover crop suppression practices and natural enemies of armyworm (Lepidoptera: Noctuidae) in no-till corn. *Environmental Entomology*, **21(1):** 41–49.

Lawrence, I.F. and Britton, E.B. 1991. Coleoptera. *In:* The insect of Australis 2nd ed. Melbourne University Press, Melbourne. 543–583 pp.

Lecordier, C. 1988. Le genre *Hyparpalus* Alluaud en Afrique (Coleoptera: Carabidae, Harpalinae). *Annales de la Société Entomologique de France* (N.S.), **24:** 1–38. Paris.

Lee, J.C. and Edwards, D.L. 1999. Impact of predatory Carabids on below and above ground pests and yield in strawberry. *Biocontrol*, **57:** 515–522.

Leisiewicz, D.S., Leisiewicz, J.L., Jr. Bradley, J.R. and Van Duyn, J.W. 1982. Serological determination of carabid (Coleoptera: Adephaga) Predation of corn earworm (Lepidoptera: Noctuidae) in field corn. *Environmental Entomology*, **11:** 1183–1186.

Leng, C.W. 1920. Catalogue of the Coleoptera of America, North of Mexico. John D. Sherman, Jr., Mt. Vernon, New York. 470 p.

Lenski, R.E. 1984. Food limitation and competition: A field experiment with two Carabus species. *Journal of Animal Ecology*, **53:** 203–216.

Li, B.C., Liang, J.E., He, J.X. and Lin, Y.Y. 1983. Effect of a mixture of BHC and parathion on *Nilaparvata lugens* Stål and its influence on natural enemies in paddy field. *Natural Enemies of Insects*, **5(1):** 20–26.

Liebke, M. 1931. Die afrikanischen Arten der Gattung Colliuris Degeer (Col. Car.). *Revue de Zoologie et de Botanique Africaines*, **20:** 280–301. Bruxelles.

Liebke, M. 1934. Die brachyninae des Afrikanischen Festlands. *Mémoires de Société Royale d'entomologie de Belgique*, **24:** 5–94. Bruxelles.

Liebman, M. and Gallandt, E.R. 1997. Many little hammers: Ecological approaches for management of crop-weed interactions. *In:* L.E. Jackson (ed.). Ecology in Agriculture. San Diego, Academic Press.

Lim, G.T. and Pan, Y.C. 1980. Entomofauna of Sugarcane in Malaysia. *Entomolgy*, 1658–1679.

Lindroth, C.H. 1954. Die Larve von *Lebia chlorocephala* Hoffm. (Col., Carabidae). *Opuscula Entomologica*, **19:** 29–32.

Lindroth, C.H. 1955. The carabid beetles of Newfoundland. *Opuscula Entomologica, Supplementum*, **12:** 1–160.

Lindroth, C.H. 1957. The Faunal Connections between Europe and North America. Wiley, New York, NY.

Lindroth, C.H. 1961. The ground beetles (Carabidae, excl. Cicindelinae) of Canada and Alaska. Part 2. *Opuscula Entomologica Supplementum*, **20:** 1–200.

Lindroth, C.H. 1961–69. The ground beetles (Carabidae, excl. Cicindelidae) of Canada and Alaska. *Opuscula Entomologica Supplementum*, **20, 24, 29, 33, 34, 35:** 1–1192.

Lindroth, C.H. 1963. The ground beetles (Carabidae, excl. Cicindelinae) of Canada and Alaska. Part 3. *Opuscula Entomologica Supplementum*, **24:** 201–408.

Lindroth, C.H. 1966. The ground beetles (Carabidae, excl. Cicindelinae) of Canada and Alaska. Part 4. *Opuscula Entomologica Supplementum*, **29:** 409–648.

Lindroth, C.H. 1985. The Carabidae (Coleoptera) of Fennoscandia and Denmark. *Fauna Entomologica Scandinavica*, **15(1):** 1–205.

Lindroth, C.H. 1972. Taxonomic notes on certain British ground-beetles (Col., Carabidae). *Entomologist's Monthly Magazine*, **107:** 209–223.

Lindroth, C.H. 1974. Coleoptera: Carabidae. Handbooks for the Identification of British Insects, **4(2):** 1–148. London: Royal Entomological Society.

Lindroth, C.H. 1985–86. The Carabidae (Coleoptera) of Fennoscandia and Denmark. *Fauna Entomologica Scandinavica*, **15:** 1. E.J. Brill, Leiden.

Löbl, I. and Smetana, A. 2003. Catalogue of Palaertic Coleoptera. Volume I. Archostemata- Myxophaga-Adephaga. Apollo Books Stenstrup, Denmark. 819 p.

Loganathan, J. and David, P.M.M. 1999. Predator complex of the teak defoliator, *Hyblaea puera* Cramer (Lepidoptera: Hyblaeidae) in an intensively managed teak plantation at Veeravanallur, Tamil Nadu. *Entomon.*, **24(3):** 259–263.

Loreau, M. 1984a. Composition et structure de trois peuplements forestier de Carabides. *Academie Royale de Belgiqu Bulletin de la Classe des Sciences*, **70:** 125–160.

Loreau, M. 1984c. Population density and biomass of Carabidae (Coleoptera in forest community. *Pedobiologia*, **27:** 269–278.

Lorenz, W. 2005. Systematic list of Extant Ground Beetles of the World (Insecta Coleoptera "Geadephaga": Trachypachidae and Carabidae incl. Paussinae, Cicindelinae, Rhysodinae). Second edition. Published by the author, Hörmannstrasse 4, D-82327 Tutzing, Germany. 530 p.

Lorenz, W. 2005b. Nomina Carabidarum: A directory of the scientific names of ground beetle (Insecta, Coleoptera " Geadephaga": Trachypachidae and Carabidae incl. Paussinae, Cicindelinae, Rhysodinae). Published by W. Lorenz, Tutzing, Germany, 2nd Edition. 993 p.

Loughridge, A.H. and Luff, M.L. 1983. Aphid predation by *Harpalus rufipes* (De Geer) (Coleopteran: Carabidae) in the laboratory and field. *Journal of Applied Ecology*, **20:** 451–462.

Lövei, G. 2008. Ecology and conservation biology of ground beetles (Coleoptera: Carabidae) in an age of increasing human dominancy. http://real-d.mtak.hu/121/1/Lovei.pdf. (Accessed on 15th Augut 2018)

Lovei, G.L., Jorgensen, H.B. and Mc Cambridge, M. 2000. Effect of proteinase inhibitor across trophic levels: Short vs. long-term consequences for a predator. *Antenna*, London, **245:** 78.

Lövei, G.L. and McCambridge, M. 2002. Adult mortality and minimum lifespan of the ground beetle *Harpalus affinis* (Coleoptera: Carabidae) in New Zealand. *New Zealand Journal of Zoology*, **29:** 1–4.

Lovei, G.L. and Sarospataki, M. 1990. Carabid beetles in agricultural fields in eastern Europe. *In:* N.E. Stork (ed.). The Role of Ground Beetles in Ecological and Environmental Studies Intercept. Andover.

Lövei, G.L. and Sunderland, K.D. 1996. Ecology and behavior of ground beetles (Coleoptera: Carabidae). *Annual Review of Entomology*, **41:** 231–256.

Luck, R.F., Shepard, B.M. and Kenmore, P.E. 1988. Experimental methods for evaluating arthropod natural enemies. *Annual Review of Entomology*, **41:** 241–246.

Luff, M.L. 1966. The abundance and diversity of the beetle fauna of grass tussocks. *Journal of Animal Ecology*, **35:** 189–208.

Luff, M.L. 1973. The annual activity pattern and life cycle of *Pterosticbus madidus* (F.) (Col. Carabidae). *Entomologica Scandinavica*, **4:** 259–273.

Luff, M.L. 1980. The biology of the ground beetle *Harpalus rufipes* in a strawberry field in Northumberland. *Annals of Applied Biology*, **94(2):** 153–164.

Luff, M.L. 1981. Notes on the identification of some Carabidae – 2. *Coleopterists' Newsletter*, **6:** 2–3.

Luff, M.L. 1982. Population dynamics of Carabidae. *Annals of Applied Biology*, **101:** 164–170.

Luff, M.L. 1987. Biology of polyphagous ground beetles in agriculture. *Agricultural Zoology Reviews*, **2:** 237–238.

Luff, M.L. 1989. Further new Coleoptera from the Isle of Man, 1987. *Entomologist's Monthly Magazine*, **125:** 118.

Luff, M.L. 1990. *Pterostichus rhaeticus* Heer (Col., Carabidae), a British species previously confused with *P. nigrita* (Paykull). *The Entomologst's Monthly Magazine*, **126:** 245–249.

Luff, M.L. 1996a. Further additions to the Manx list of Coleoptera. *Entomologist's Monthly Magazine*, **132:** 29–31.

Luff, M.L. 2002. Carabid assemblage organization and species composition. *In:* J.M. Holland (ed.). The Agroecology of Carabid Beetles. Intercept, Andover, UK.

Luff, M.L., Eyre, M.D. and Rushton, S.P. 1992. Classification and prediction of grassland habitats using ground beetles (Coleoptera, Carabidae). *Journal of Environmental Management*, **35:** 301–315.

Luff, M.L. and Wardle, J. 1991. *Trechus rivularis* (Gyll.) (Col, Carabidae) in Northumberland. *Entomologist's Monthly Magazine*, **127:** 42.

Loughridge, A.H. and Luff, M.L. 1983. Aphid predation by *Harpalus rufipes* (Degeer) (Coleoptera:Carabidae) in the laboratory and field. *Journal of Applied Ecology*, **20:** 451–462.

Luka, H., Lutz, M., Blick, T. and Pfiffner, L. 2001. Einfluss cingesäter Wildblumenstreifen auf die epigäischen Laukäfer und Spinnen (Carabidae und Araneae) in der intensiv genutzten Agrarlandschaft Grosses Moos. *Schweiz Peckiana*, **1:** 45–60.

Lund, R.D. and Turpin, F.T. 1977a. Serological investigation of Black Cutworm larval consumption by ground beetles. *Annals of The Entomological Society of America*, **70:** 322–324.

Lund, R.D. and Turpin, F.T. 1977b. Carabid damage to weed seeds found in Indiana cornfields. *Environmental Entomology*, **6:** 695–698.

Lundgren, J.G. 2005. Ground beetles as weed control agents: Effects of farm management on granivory. *American Entomology*, **51:** 224–226.

Lutz, H. 1990. Systematische und paloekologische Untersuchungen an insekten aus dem mittel – eozaen der Grube Messel bei Darmstadt. *Courier Forchungsinstitute Senkenberg*, **124:** 1–165.

Lys, J.A. and Nentwig, W. 1991. Surface activity of carabid beetles inhabiting cereal fields: Seasonal phenology and the influence of farming operations on five abundant species. *Pedobiologia*, **35(3):** 129–138.

Lys, J.A. 1994. The positive influence of strip management on ground beetles in a cereal field: Increase, migration and overwintering. *In:* K. Desender, M. Dufrene, M. Leraau, M.L. Cuff and J.P. Maelfait (eds). Carabid Beetles: Ecology and Environment. Kluwer Academic, Dordrecht.

Maclean, B.D. and Usis, J.D. 1992. Ground Beetles (Coleoptera: Carabidae) of Eastern Ohio Forests threatened by the Gypsy Moth, *Lymantria dispar* (L.) (Lepidoptera: Lymantridae). *Ohio Journal of Science*, **92(3):** 46–50.

Macrae, T. 2011. Diversity in Tiger Beetle Larval Burrows (On-line). Beetles in the Bush. Accessed March 28, 2013 at http://beetlesinthebush.wordpress.com/2011/01/13/diversity-in-tiger-beetle-larval-burrows/ (Accessed on August 14, 2018)

Maddison, D.R., Baker, M.D. and Ober, K.A. 1999. Phylogeny of carabid beetle as inferred from 18s ribosomal DNA (Coleoptera: Carabidae). *Systematic Entomology*, **24:** 103–138.

Maddison, D.R. and Ober, K.A. 2000. Phylogeny of minute carabid beetles and their relatives based upon DNA sequence data (Coleoptera, Carabidae, Trechitae). *Zookeys*, **147:** 229–260.

Madge, Ronald Bradley. 1967. A revision of the genus Lebia Latreille in America north of Mexico (Coleoptera, Carabidae). *Quaestiones Entomologicae*, **3:** 139–242.

Mahammad, S.U. and Jahan, M. 2004. Schedule and Need Based Chemical control of Brown Planthopper and their impact on predator *Ophionea indica* Thunberg. *Asian Journal of Plant Sciences*, **3(60):** 687–689.

Makarov, K.V. 1994. Annual reproduction rhythms of ground beetles: A new approach to the old problem. *In:* Desender, K., Dufrene, M., Loreau, M., Cuff, M.L.and Maelfaitjp (eds). Carabid Beetles: Ecology and Evolution. Kluwer Academic Dodrecht.

Mani, M.S. 1990. General Entomology (Third edition) Oxford and IBH. 912 p.

Manjunath, T.M., Rai, P.S. and Gowda, G. 1978. Natural enemies of brown planthopper and green leafhopper in India. *International Rice Research Notes*, **3(2):** 11.

McCarty, M.T., Shepard, M. and Turnipseed, S.G. 1980. Identification of predaceous arthropods in soybeans by using autoradiography. *Environmental Entomology*, **9:** 199–203.

McWhorter, R.E., Grant, J.F. and Shepard, M. 1984. Life history of a predator, *Calleida decora*, and the influence of temperature on development. *Journal of Agricultural Entomology*, **1:** 68–77.

Mateu, J. 1966. Coléoptéres Carabiques récoltés par J. Mateu dans I' Ennedi et au Nord – Tschad. *Bulletin de l'Institut Français d'Afrique Noire*, Ser. A, **28:** 1501–1543. Dakur.

Menalled, F.D., Lee, J.C. and Landis, D.A. 2001. Herbaceous filter strips in agroecosystems: Implications for ground beetle (Coleoptera: Carabidae) conservation and invertebrate weed seed predation. *Great Lakes Entomologist*, **38(2):** 472–483.

Meszaros, Z. (ed.) 1984a. Results of faunistical and forestical studies in Hungarian apple orchards. *Acta Phytopathologica Academiae Scientiarum Hungaricae*, **19:** 91–176.

Metalin, A.V. 1992. Correlation of the crawling and flying migration in populatios of the dominant species of the carabid beetles (Insecta, Coleoptera, Carabidae) in the south-west of the steppe zone. *Zoologicheskii Zhurna*, **71:** 57–68 (in Russian).

Misra, R.M. 1975. Note on *Anthia sexguttata* F. (Carabidae: Coleoptera) a new predator of *Pyrausta machaeralis* Walk. and *Hyblaea puera* Cramer. *Indian Forester*, **101:** 605.

Mitchell, B. 1963. Ecology of two carabid beetles, *Bembidion lampros* (Herbst) and *Trechus quadristriatus* (Scharnk). *Journal of Animal Ecology*, **32(3):** 377–392.

Mohamed Jalaluddin, S. 1999. Integrated pest management of guava fruit fiy. *Indian Farming*, **48(10):** 14.

Mohamed, V.K.V., Abdurahiman, V.C. and Ramadevi, O.K. 1982. Notes on two carabid predators (Coleoptera: Carabidae) on *Nephantis serinopa* (Xylorictidae: Lepidoptera) from Kerala. *Entomon*, **7:** 341–343.

Mols, P.J.M. 1988. Simulation of hunger, feeding and egg production in the carabid beetle *Pterostichus coerulescens* L. *Agricultural University Wageningen Papers*, **88(3):** 1–99.

Moore, B.P. 1956. A new name for *Tachys piceus* Edmonds (Col. Carabidae). *Entomologist's Gazette*, **7:** 87–88.

Moore, B.P. 1972. Description of the larva of *Siagon*a (Coleoptera: Carabidae). *Journal of Entomology* (B), **41(2):** 155–157.

Morrill,W.L. 1975. Plastic pitfall trap. *Environmental Entomology*, **4:** 596.

Najmeh S. and Hamid S. 2014. A study on the ground beetles in Mazandaran province, northern Iran. *Entomofauna Zeitscrift Für Entomologie Band* 36 Heft, **37:** 505–512.

Nash, D.R. 1979. *Bradycellus csikii* Lacz. (Col., Carabidae) discovered in Suffolk. *Entomologist's Record*, **91:** 279–280.

Nash, D.R. 1983. Notes concerning the habitat of, and other Coleoptera associated with, an example of *Agonum gracilipes* (Duft.) (Col.:

Carabidae) together with a reappraisal of its British status. *Entomologist's Record*, **95:** 205–206.

Naviaux, R. 1995. Les Collyris (Coleoptera, Cicindelidae). Révision des genres et description de nouveaux taxons. Société Linnéenne de Lyon.

Neal, T.M. 1974. Predaceous Arthropods in the Florida Soybean Agroecosystem. M.S. Thesis University of Florida. 194 p.

Neculiseanu, Z.Z. and Matalin, A.V. 2000. A Catalogue of the Ground-Beetles of the Republic of the Moldova (Insccta, Coleoptera: Carabidae). Pensoft Publish., Sofia-Moskow. 164 p.

Nelemans, M.N.E. 1987a. On the life history of the carabid beetle *Nebria brevicollis* (F.). *Netherlands Journal of Zoology*, **37:** 26–42.

Nelemans, M.N.E. 1987b. Possibilities for flight in the carabid beetle *Nebria brevicollis* (F.). The impotance of food during the larval growth. *Oecologia*, **72:** 502–509.

Nelemans, M.N.E., den Boer, P.J. and Spee, A. 1989. Recruitment and summer diapauses in the dynamics of a population of *Nebria brevicollis* (Coleoptera: Carabidae). *Oikos*, **56:** 157–169.

Niemela, J., Halme, E. and Haila, Y. 1990. Balancing sampling effort in pitfall trapping of carabid beetle. *Entomologica Fennica*, **1:** 232–238.

Öncüer, C.1991. A Catalogues of the Parasites and Predators of Insect Pest of Turkey. Ege University, Faculty of Agriculture Publication, İzmir. 354 p.

Paarmann, W. 1979. A reduced number of larval instars, as an adaptation of the desert carabid beetle *Thermophilum* (*Anthia*) *sexmaculatum* F. (Coleoptera, Carabidae) to its arid environment. On the Evolution of Behaviour in Carabid Beetles. P.J. Den Boer, H.U. Thiele and F. Weber (eds). Veenman and Zonen B.V. *Miscellaneous Papers, Agricultural University, Wageningen, The Netherlands* **18:** 113–117.

Paarrmann, W. 1986. Seasonality and its control by environmental factors in tropical ground beetles (Col. Carabidae) *In:* P.J. den Boer, M.L. Luff, D. Mossakowski, F. Weber (eds). Carabid Beetles. Their Adaptation and Dynamics. Fischer Verlag, Stuttgard/New York.

Pakarinen, E. 1994. The importance of mucus as a defence against carabid beetles by the slugs *Asion fasciaitus* and *Deroceras reticulatum*. *Journal of Molluscan Studies*, **60:** 149–155.

Pallet, M. and Desender, K. 1987. Feeding ecology of grassland-inhibiting carabid beetles (Carabidae, Coleoptera) in relation to the availability of some prey groups. *Acta Phytopathologica et Entomologica Hungarica*, **22:** 223–246.

Palmer, M.W. 1993. Putting things in even better order: The advantages of canonical correspondence analysis. *Ecology*, **74:** 2215–2230.

Park, J.K., Trac, D.H. and Will, K. 2006. Carabidae from Vietnam (Coleoptera). *Journal of Asia-Pacific Entomology*, **9(2):** 85–105.

Parker, J.R. and Wakeland Claude. 1957. Grasshopper egg pods destroyed by larvae of bee flies, blister beetles and ground beetles. *U.S. Department of Agriculture Technical Bulletin*, **1165:** 29.

Parman, Θistein. 1979. Marcello Haugen. J.W. Cappelens folag, Oslo.

Parmenter, R.R and MacMahon, J.A. 1988. Factors influencing species composition in a ground beetle community (Carabidae): Predation by rodents. *Oikos.*, **52:** 350–356.

Parry, J.A. 1975. *Dyschirius angustatus* Ahrens (Col., Carabidae) in Fast Sussex. *Entomologist's Monthly Magazine*, **111:** 160.

Patil, V.J. and Sathe, T.V. 2003. Insect Predators and Pest Management. Daya Publishing House, New Delhi. 209 p.

Pearson, D., Knisley, C. and Kazilek, C. 2006. A Field Guide to the Tiger Beetles of the United States and Canada: Identification, Natural History, and Distribution of the Cicindelidea. Oxford Univeristy Press, New York.

Perera, P.A.C.R. 1981. Predation studies on *Pareuchaetes pseudoinsulata* (Lep: Arctiidae) using Labelled Immatures. *Ceylon Coconut*, **32:** 105–110.

Pillai, G.B. and Nair, K.R.1990. On the biology of *Calleida splendidula* (F.) (Coleoptera: Carabidae), a predator of the coconut leaf eating caterpillar, *Opisina arenosella* Wlk. *Indian Coconut Journal*, **20(12):** 14–17.

Pitre, H.N. and Chapman, R.K. 1964. Ground beetles as predators of cabbage maggot eggs. *Entomological Society of America, North Central Branch; Proceeding*, **19:** 102–103.

Poetker, E. 2003. *Brachinus fumans*. (On-line) Animal Diversity Web. Accessed 2 November 2007 at http://animaldiversity.ummz.umich.edu/site/accounts/information/Brachinus_fumans. html (Accessed on July 12, 2017)

Poinar, G.O. 1975. Entomogenous Nematodes: A Manual and Host List of Insect-Nematode Associations. Brill Leiden. 317 p.

Ponomarenko, A.G. 1977. Mesozoic Coleoptera. *Trans Paleontological Institute Moscow*, **161:** 1–204.

Ponomarenko, A.V. 1969. The ground beetle *Anisodactylus signatus* (Coleoptera, Carabidae) – maize pest in Rostov district. *Zoologicheskii Zhurnal*, **48:** 143–146 (in Russian).

Poprawski, T.J. 1994. Insect parasites and predators of Phyllophaga anxia (LeConte) (Col., Scarabaeidae) in Quebec, Canada. *Journal of Applied Entomology*, **117:** 1-9.

Potts, G.R. and Vickerman, G.P. 1974. Studies on the cereal ecosystem. *Advances in Ecological Research*, **8:** 107–197.

Poulin, G. and O'Neil, L.C. 1969. Observations sur les predatèurs de la limace noire, *Arion ater* (L.) (Gastèropodes, Pulmones, Arionidès). *Phytoprotection*, **50:** 1–6.

Powell, W., Dean, G.J. and Dewar, A. 1985. The influence of weeds on polyphagous arthropod predators in winter wheat. *Crop Protection*, **4:** 298–312.

Prasad, K. and Rajagopal, D. 1990. Carabid beetle, *Omphra pilosa* Klug (Coleoptera: Carabidae): A potential predator on termites. *Journal of Biological Control*, **4(2):** 106–109.

Proctor, W. 1946. Biological survey of the Mount Desert Region. Part 7. The insect fauna. Biological Survey of the Mount Desert Region, Inc., Cornfield, Bar Harbor, Maine. 566 p.

Purvis, G., Fadl, A. and Bolger, T. 2001. A multivariate analysis of cropping effects on Irish ground beetle assemblages (Coleoptera: Carabidae) in mixed arable and grass farmland. *Annals of Applied Biology*, **139:** 351–360.

Pushpalatha, N.A. and Veeresh, G.K. 1995. Numerical relationship between *Opisina arenosella* Walker and its natural enemies, *Apanteles taragamae* Vierick and *Parena nigrolineata* Chaudoir on coconut. *Journal of Insect Science*, **8(2):** 148–150.

Puttarudriah, M. and Raju, R.N. 1952. Observations on the host range, and control of *Azazia rubricans*. *Indian Journal Entomology*, **14(2):** 158.

Rai, B.K., Joshi, H.C., Rathore, Y.K., Dutta, S.M. and Shinde, V.K.R. 1969. Studies on the bionomics and control of white grub *Holotrichia consanguinea* Blanch in Lalsot, Dist. Jaipur, Rajasthan. *Indian Journal of Entomology*, **31(2):** 132–142.

Raj, T.S., Sabu, K.T. and Danyang, K. 2012. The apterous endemic genus *Omphra* Dejean (Coleoptera: Carabidae; Helluonini) of the Indian subcontinent: Taxonomy with notes on habits and distribution pattern. *Insect Mundi*. Center for Systematic Entomology, Inc. USA (http//www. center for systematicentomology.org/) (Accessed on October 17, 2017)

Rajagopal, D. and Kumar, P. 1988. Predation potentiality of *Chlaenius panagaeoides* Chaudoir on cowpea aphid. *Journal of Aphidology*, **2:** 93–99.

Rajagopal, D. and Kumar, P. 1992. Carabids (Coleoptera: Carabidae) as potential predators on major crop pests in South India. *Journal of Biological Control*, **6(1):** 13–17.

Ranjha, M. and Irmler, U. 2014. Movement of carabids from grassy strips to crop land in organic agriculture. *Journal of Insect Conservation*, **18:** 457–467.

Rao, V.P., Ghani, M.A., Sankaran, T. and Mathur, K.C. 1971. A review of the biological control of insects and other pests in south east Asia and the pacific region. *Technical Bulletin of the Commonwealth Institute of Biological Control*, **6:** 149.

Rao, A.G. 1978. Preliminary observation on the biology of *Parena laticincta* Bates (Coleoptera: Carabidae). A predator of *Nephantis serinopa* Meyrick. *Indian Coconut Journal*, **9(1):** 2-5.

Raworth, D.A., Clements, S.J., Cirkony, C. and Bousquet, Y. 1997. Carabid beetles in commercial raspberry fields in the Fraser Valley of British Columbia and a sampling protocol for *Pterostichus melanarius*

(Coleoptera: Carabidae). *Journal of the Entomological Society of British Columbia*, **94:** 51–58.

Reichardt, H. 1971. Notes on the bombarding behavior of three carabid beetles (Coleoptera, Carabidae). *Revista Brasileira de Entomologia*, **15:** 31–34.

Ribera, I., McCracken, D. and Luff, M.L. 1996. *Agonum viduum* (Panzer) and *A. moestum* (Duftschmid) (Carabidae) in Scotland. *Coleopterist*, **5:** 56–57.

Richman, D.B., Hemenway, R.C. Jr. and Whitcomb, W.H. 1980. Field cage evaluation of predators of the soybean looper, *Pseudoplusia includens* (Lepidoptera: Noctuidae). *Environmental Entomology*, **9:** 315–317.

Riddick, E.W. 2008. Ground Beetle (Coleoptera: Carabidae) Feeding Ecology. *In:* J.L. Capinera (ed.). *Encyclopedia of Entomology,* **4:** 1742–1747. Springer Science, Germany.

Riddick, E.W. and Mills, N.J. 1994. Potential of adult carabids (Coleoptera:Carabidae) as predator of fifth instar codling moth (Lepidoptera: Tortricidae) in apple orchards in California. *Environmental. Entomology*, **23:** 1338–1345.

Rijnsdorp, A.D. 1980. Pattern of movement in and dispersal from a Dutch forest of *Carabus problematicus* Hbst (Coleoptera, Carabidae). *Oecologia*, **45:** 274–281.

Rivard, I. 1964a. Carabid beetles (Coleoptera: Carabidae) from agricultural lands near Belleville, Ontario. *Canadian Entomologist*, **96:** 517–520.

Rivard, I. 1964b. Observations on the breeding periods of some ground beetles (Coleoptera: Carabidae) in eastern Ontario. *Canadian Journal of Zoology*, **42:** 1081–1084.

Rivard, I. 1965. Addition to the list of carabid beetles (Coleoptera: Carabidae) from agricultural lands near Belleville, Ontario. *Canadian Entomologist*, **97:** 332–333.

Rivard, I. 1966. Ground beetles (Coleoptera: Carabidae) in relation to agricultural crops. *Canadian Entomologist*, **98:** 189–195.

Rivard, I. 1974. Faune carabique d'un verger expérimental. *Phytoprotection*, **55:** 55–63.

Roff, D.A. 1994. The evolution of flightlessness: Is history important? *Evolutionary Ecology*, **8:** 639–657.

Ruiter, P.C., van Stralen, M.R., van Euwijk, F.A., Slob, W., Bedaux, J.J.M. and Ernsting, G. 1989. Effect of hunger and prey traces on search activity of the predatory beetle *Notiophilus biguttatus*. *Entomologia Experimentalis et Applicata*, **51:** 87–95.

Saha, S.K. 1986. Ground beetles (Insecta, Coleoptera: Carabidae) of Silent Valley (Kerala, India) *Records of the Zoological Survey of India*, **84(1-4):** 67–77.

Saha, S.K. and Halder, S.K. 1986. Tiger beetles (Coleoptera, Cicindelidae) of silent valley (Kerala, India). *Records of the Zoological Survey of India*, **84(1–4):** 131–142.

Saha, S.K., Mukherjee, A.K. and Sengupta, T. 1992. *Records of the Zoological Survey of India*, Occasional Paper No 144, 3 Carabidae (Coleoptera: Insecta) of Calcutta. ZSI, New Alipore, Kolkata.

Sahlberg, J. 1912–1913. Coleoptera mediterranea orientalis, quae in Aegypto, Palaestina, Syria, Caramania, atque in *Anatolia occidentalia* anno 1904. *Öfversigt at Finska Vetenskaps – Societetens Förhandlingar*, Bd. LV,Afd, A. No. 1.

Saipulaeva, B.N. 1986. Characteristics of the habitat distribution of geobiont beetles (Coleoptera: Carabidae, Scaraboeidae, Elateridae, Tenebrionidae) of the Irganaiskaya depression in the central mountains of Deghestan. *Obzor Entomology*, **65:** 96–106 (in Russian).

Samal, P. and Mishra, B.C. 1984. *Ophionea indica*, a predatory carabid beetle of rice brown planthopper *Nilaparvata lugens* (Stål). *Oryza*, **19:** 212.

Santos, B.S. 2014. Distribution patterns of tiger beetle species in the Philippines and Southeast Asia. *Journal of Entomology and Zoology Studies*, **2(4):** 271–275.

Saraswati, K.C. 1990. Natural enemies of *Rhopalosiphum nymphaeae* L. (Homoptera: Aphididae) infesting *Euryale ferox* Salisbury in North Bihar. *Newsletter Aphidological Society of India*, **1:** 11–13.

Saska, P. and Honek, L. 2004. Development of the beetle parasitoids, *Brachinus explodes* and *B. crepitans* (Coleoptera: Carabidae). *Journal of Zoology*, **262(1):** 29–36.

Saska, P., Vodde, M., Heijerman, T., Westerman, P. and Werf, W. 2007. The significance of a grassy field boundary for the special distribution of carabids within two cereal fields. *Agriculture Ecosystems and Environment*, **122:** 427–434.

Satpathi, C.R. 2000. Preys of the Carabid Beetle *Anthia sexguttata* and Tiger beetle *Cicindela* sp. *Insect Environment*, **6(1):** 16.

Satpathi, C.R., Maiti, A.K. and Samanta, A. 2000. Survey of predators controlling insect pests of different crops in the Eastern Himalayan region. *Indian Journal of Environment and Ecoplanning*, **3(1):** 167–171.

Sawada, H. 2001. Further new records of Tiger Beetle species from China (Coleopter: Cicicindelidae). *Entomological Review of Japan*, **56(1):** 23–24.

Schacht, W. 2000. Insekten aus Gambia, Westafrika (Diptera; Platypezidae, Odiniidae, Tabanidae, Glossinidae und Planipennia; Chrysopidae, Myrmeleontidae, Ascalaphidae sowie, Coleoptera: Carabidae, Cicindelidae, Elateridae, Scarabaeidae). *Entmophona*, **21(1):** 1–4. Ansfelden.

Scheller, H.V. 1984. The role of ground beetles (Carabidae) as predators on early populations of cereal aphids in spring barley. *Zeitschrift für Angewandte Entomologie*, **97:** 451–463.

Schmaedick, M.A. and Shelton, A.M. 2000. Arthropod predators in cabbage (Cruciferae) and their potential as naturally occurring biological control agent *Pieris rapae* (Lepidoptera: pieridae). *The Canadian Entomologist*, **132:** 655–675.

Serrano, J. and Lencina, J.L. 2009. *Harpalus* (*Baeticoharpalus*) *lopezi*, new subgenus and new species of *Harpalus* from Iberian Peninsula (Coleoptera, Carabidae: Harpalini). *Entomologica Fennica*, **19:** 193–198.

Shanower, T.G. and Ranga, G.V.R. 1990. *Chlaenius* sp. (Coleoptera: Carabidae): A predator of groundnut leaf miner larvae. *International Arachis Newsletter*, **8:** 19-20.

Sharma, A.K. and Bisen, U.K. 2013. Taxonomic documentation of insect pest fauna of vegetable ecosystem collected in light trap. *International Journal of Environmental Science: Development and Monitoring*, **4(3):** 4–10.

Sharp, D. 1913. *Bradycellus distinctus* Dej., in England. *Entomologist's Monthly Magazine*, **49:** 54.

Shearin, A.F., Reberg-Horton, S.C. and Gallandt, E.R. 2007. Direct effects of tillage on the activity density of ground beetle (Coleoptera: Carabidae) weed seed predators. *Environmental Entomology,* **36:** 1140–1146.

Shelford, Victor E. 1963 The Ecology of North America. University of Illinois Press, Urbana, III. 610 p.

Shepard, B.M., Barrion, A.T. and Litsinger, J.A. (1988). Helpful Insects, Spiders and Pathogen. International Rice Research Institute, Philippines.

Shepard, B.M., Carner, G.R., Barrion, A.T., Ooi, P.A.C. and van den Berg, H. 1999. Insects and Their Natural Enemies Associated with Vegetables and Soybean in Southeast Asia. Quality Printing Company, Orangeburg, SC.108.

Siemann, E., 1998. Experimental tests of effects of plant productivity and diversity on grassland arthropod diversity. *Ecology*, **79:** 2057–2070.

Sigsgaard, L., Villareal, S., Gapud, V. and Rajotte, E. 1999. Directional movement of predators between the irrigated rice field and its surroundings. *In:* L.W. Hong and S.S. Sastroutomo (eds). Symposium on Biological Control in the Tropics. CABI Publishing, Wallingford, UK.

Skuhravy, V. 1959. Die Nährung der Feldcarabiden. *Acta Societatis Entomologicae Cechosloveniae*, **56:** 1–18.

Slough, W.W. 1940. The feeding of ground beetles (Carabidae). *American Midland Naturalist*, **24:** 336–344

Sopp, P. and Wratten, S.D. 1986. Rates of consumption of cereal aphids by some polyphagous predators in the laboratory. *Entomologia Experimentalis et Applicata*, **41:** 69–73.

Sota, T. 1984. Long adult life span and polyphagy of a carabid beetle, *Leptocarbus kumagaii* in relation to reproduction and survival. *Researches on Population Ecology*, **26:** 389–400.

Sota, T. 1985. Limitation of reproduction by feeding condition in a carabid beetle *Carabus yaconius Researches on Population Ecology*, **27:** 171–184.

Sota, T. 1987. Mortality pattern and age structure in two carabid populations with different seasonal life cycles. *Researches on Population Ecology*, **29:** 237–254.

Southwood, T.R.E and Comins, H.N. 1976. A synoptic population model. *Journal of Animal Ecology*, **45:** 949–965.

Speight, M.C.D. 1976b. *Badister meridionalis*, *Megasyrpbus annulipes* and *Wesmaelius quadrifasciatus*: Insects new to Ireland. *Irish Naturalist's Journal*, **18:** 303–305.

Speight, M.C.D. 1977. The ground beetles *Dyscbirius luedersi* new to Ireland, *Badister peltatus* and *Chlaenius tristis* re-instated as Irish. *Irish Naturalist's Journal*, **19:** 116–118.

Speight, M.C.D., Anderson, R. and Luff, M.L. 1982. An annotated list of the Irish ground beetles (Col., Carabidae + Cicindelidae). *Bulletin of the Irish Biogeographical Society*, **6:** 25–53.

Speight, M.C.D., Martinez, M. and Luff, M.L. 1986. The *Asaphidion* (Col.: Carabidae) species occurring in Great Britain and Ireland. *Proceedings and Transactions of the British Entomological and Natural History Society*, **19:** 17–21.

Spence, J.R. and Spence, D.H. 1988. Of ground-beetles and men: Introduced species and the synanthropic fauna of western Canada. *Memoirs of the Entomological Society Canada*, **144:** 151–168.

Stork, N.E. 1987. Adaptations of arborial carabids to life in trees. *Acta Phytopathologica et Entomologica Hungarica*, **22(1-4):** 273–291.

Stork, N.D. 1990. The Role of Ground Beetles in Ecological and Environmental Studies. Andover: Intercept. 424 p.

Straneo, S.L.1939. On some new species of African Pterostichini (Col. Carab.). Pt. I, II. *Proceedings of the Royal Entomological Society* London (B), **8:** 167–174; 175–180. London.

Straneo, S.L.1950. Revisione delle specie africane del gen, *Melanchiton* Andrewes (*Melanodes* Chaudoir et Auctt.) (Col. Carab.). *Revue de Zoologie Africaine*, **44:** 61–104. Bruxelles.

Straneo, S.L. 1956b. Coléoptéres recueillis par N. Leleup au lac Tumba IV. Coleoptera Carabidae Pterostichinae. *Revue de Zoologie Africaine*, **54:** 127–136. Bruxelles.

Straneo, S.L.1956c. Su alcuni Pterostichini entrati recentemente nel Museo del Congo belga (Coleoptera Carabidae). *Revue de Zoologie Africaine*, **54:** 257–275. Bruxelles.

Suenaga, H. and Hamamura, T. 1998. Laboratory evaluation of carabid beetles (Coleoptera, Carabidae) as predators of diamondback moth (Lepidoptera: Plutellidae) larvae. *Environmental Entomology*, **27(3):** 267–272.

Sunderland, K.D. 1975. The diet of some predatory arthropods in cereal crops. *Journal of Applied Ecology*, **12**: 507–515.

Sunderland, K.D., Lovei, G.L. and Fenlon, J. 1995. Diets and reproductive phonologies of the introduced ground beetles *Harpalus affinis* and

Clivina australasiae (Coleoptera: Carabidae) in New Zealand. *Australian Journal of Zoology*, **43**: 39–50.

Sunderland, K.D. and Vickerman, G.P. 1980. Aphid feeding by some polyphagous predators in relation to aphid density in cereal fields. *Journal of Applied Ecology*, **17:** 389–396.

Sunderland, K.D. 2002. Invertebrate pest control by Carabids. *In:* J.M. Holland (ed.). The Agroecology of Carabid Beetles. 165–214. Andover: Intercept Ltd.

Swaminathan, R., Bhati, K.K. and Hussain, T. (2001). Preliminary investigations on the predation potential of carabids. *Indian Journal of Applied Entomology*, **15:** 37–41.

Symondson, W.O.C. and Williams, I.B. 1997. Low-vacuum electron microscopy of carabid chemoreceptors: A new tool for the identification of live and valuable museum specimens. *Entomologia Experimentalis et Applicta*, **85:** 75–82.

Takami, Y. and Sota, T. 2006. Four new species of the Australian *Pamborus* Latreille (Coleoptera, Carabidae) carabid beetles. *Australian Journal of Entomology*, **45:** 44–54.

Tamaki, G. and Olsen, D. 1977. Feeding potential of predators of *Myzus persicae* Sulz. *Journal of the Entomological Society of British Columbia*, **74:** 23–26.

Tezcan, S., Anlas, S. and Jeanne, C. 2011. Species composition and habitat selection of Ground beetles (Carabidae: Coleoptera) collected by pitfall traps in Bozadağlar Mountain, Western Turkey. *Munis Entomology and Zoology*, **6(2):** 676–685.

Thiele, H.U. 1977. Carabid Beetles in Their Environments. Springer-Verlag. Berlin, Germany. 369 p.

Thomas, M.B., Wratten, S.D. and Sotherton, N.W. 1991. Creation of island habitats in farmland to manipulate populations of beneficial arthropods; predator densities and emigration. *Journal of Applied Ecology*, **28:** 906–917.

Tischler, W. 1958. Synoekologische Untersuchungen an der Fauna der Felder und Feldgehoelze (Ein Beitrag zur Oekologie der Kulturlandschaft*).* *Zeitschrift für Morphologie und Ökologie der Tiere*, **47**: 54–114.

Toft, S. and Bilde, T. 2002. Carabid diets and food value. *In:* J.M. Holland (ed.). The Agroecology of Carabid Beetles. Intercept, Andover, UK. 81–110 p.

Tonhasca, A., Jr. 1993. Carabid beetle assemblage under diversified agroecosystems. *Entomologia Experimentalis et Applicata*, **68:** 279–285.

Tooley, J. and Brust, G.E. 2002. Weed seed predation by Carabid beetles. *In:* J.M. Holland (ed.). The Agroecology of a Carabid Beetles. Andover: Intercept Ltd. 215-229 pp.

Turin, H., Haeck, J. and Hengeveld, R. 1977. Atlas of the Carabid Beetles of the Netherlands. North Holland Publ. Co. Amsterdam.

Tyler, B.M.J. and Ellis, C. 1979. Ground beetles in three tillage plots in Ontario and observations on their importance as predators of the northern corn rootworm, *Diabrotica longicornis* (Coleoptera: Chrysomelidae). *Proceedings of the Entomological Society of Ontario*, **110:** 65–73.

Ullah Shaef Mohammad and Jahan Mahbuba (2004). Schedule and need based chemical control of brown planthopper and their impact on the predator *Ophionea indica* (Thunberg). *Asian Journal of Plant Science*, **3(60):** 687–689.

Varis, A.L., Holopainen, J.K. and Koponen, M. 1984. Abundance and seasonal occurrence of adult Carabidae (Coleoptera) in cabbage, sugar beet and timothy fields in southern Finland. *Zeitschrift für Angewandte Entomologie*, **98:** 62–73.

Vestal, A.G. 1913. An associational study of Illinois sand prairie. *Illinois State Laboratory of Natural History Bulletin*, **10:** 1–96.

Vlijm, L., van Dijk, T.S. and Wijmans, Y.S. 1968. Ecological studies on carabid beetles III. Winter mortality in adult *Calathus melanocephalus* (Linn.). Egg production and locomotory activity of the population which has hibernated. *Oecologia*, **1:** 304–314.

Vries, H.H. de and Boer, P.J. den. 1990. Survival of populations of *Agonum ericeti* Panz. (Col., Carabidae) in relation to fragmentation of habitats. *Netherlands Journal of Zoology*, **40:** 484–498.

Walkden, H.H. and Wilbur, D.A. 1944. Insects and other arthropods collected in pasture grasses, wastelands, and forage crops, Manhattan, Kansas, 1937–1940. *Kansas Entomological Society Journal*, **17(4):** 128–143.

Wallin, H. 1989. The influence of different age classes on the seasonal activity and reproduction of four medium-sized carabid species inhabiting cereal field. *Holarctic Ecology*, **12:** 201–212.

Wallin, H., Chiverton, P.A., Ekbom, B.S. and Borg, A. 1992. Diet, fecundity and egg size in some polyphagous predatory carabid beetles. *Entomologia Experimentalis et Applicata*, **65:** 129–140.

Webster, F.M. 1880. Notes on the food of predaceous beetles. *Illinois State Laboratory of Natural History Bulletin*, **1(3):** 149–152.

Weed, A.S. and Frank, J.H. 2005. Oviposition behavior of *Pheropsophus aequinoctialis* L. (Coleoptera: Carabidae): A natural enemy of *Scapteriscus* mole crickets (Orthoptera: Gryllotalpidae). *Journal of Insect Behavior*, **15:** 707–723.

Welch, R.C. 1992. *Tachys parvulus* (Dejean) (Col.: Carabidae) from synanthropic habitats in Northamptonshire and Bedfordshire. *Entomologist's Record*, **104:** 81–82.

Werling, B.P. and Gratton, C. 2008. Influence of field margins and landscape context on ground beetle diversity in Wisconsin (USA) potato fields. *Agriculture Ecosystems and Environment*, 1**28:** 104–108.

Wheater, C.P. 1989. Prey detection by some predatory Coleoptera (Carabidae and Staphylinidae) *Journal of Zoology*, **218:** 171–185.

Whelan, D.B. 1936. Coleoptera of an original prairie area in eastern Nebraska. *Kansas Entomological Society Journal*, **9:** 111–115.

Whitcomb, W.H. and Bell, K. 1964. Predaceous insects, spiders and mites of Arkansas cotton fields. *Arkansas Agricultural Experiment Station Bulletin*, **690:** 1–84.

Whitcomb, W.H., Bhatkar, A. and Nickerson, J.C. 1973. Predators of *Solenopsis invicta* queens prior to successful colony establishment. *Environmental Entomology*, **2:** 1101–1103.

Wiedenmann, R.N. and O'Neill, R.J. 1990. Effects of low rates of predation on selected life-history characteristics of *Podisus maculiventris* (Say.) (Heteroptera: Perxatomidae). *Canadian Entomologist*, **122:** 271–283.

Wiedenmann, R.N., Larrain, P.L. and O'Neil, R.J. 1992. Pitfall sampling of ground beetles (Coleoptera: Carabidae) in Indiana soybean fields. *Journal of the Kansas Entomological Society*, **65:** 279–291.

Wiesner, J. 2001. New records of tiger beetles from Gambia (Coleoptera: Cicindelidae). *Entomologische Zeitschrift*, **111(7):** 305–307.

Wilkinson, A.T.S. 1965. Releases of cinnabar moth (*Hypocrita jacobaeae* L.) on tansy ragwort in British Columbia. *Proceedings of the Entomological Society of British Columbia*, **62:** 10–12.

Will, K.W. 1998. A new species of *Diplocheila brulle* from North America, with notes on female reproductive tract characters in selected Licinini and implication for evolution of the sub genus *Isorembus jeannel* (Coleoptera: Carabidae: Licinini). *Proceedings of the Entomological Society of Washington*, **100(1):** 95–103.

Will, K.W. 2015. Resolution of taxonomic problems in Australian Harpalini, Abecetini, Pterostechini and Oodini (Coleoptera: Carabidae). *Zookeya*, **545:** 131–137.

Williams, S.A. 1984. *Cymindis macularis* (Fischer v. Waldheim) (Coleoptera: Carabidae) confirmed as a British species. *Entomologist's Monthly Magazine*, **120:** 107.

Williams, S.A. 1997. *Tachys walkerianus* Sharp (Carabidae), a patio beetle? *Coleopterist*, **6(1):** 47.

Wingo, C.W., Thomas, G.D., Clark, G.N. and Morgan, C.E. 1974. Succession and abundance of insects in pasture manure: Relation to face fly survival. http//dx.dot.org/10.1093/aesa/67.3.3.386-390 (Accessed on April 18, 2017)

Wishart, G., Doane, J.F. and Maybee, G.E. 1956. Notes on beetles as predators of eggs of *Hylemya brassicae* (Bouch), (Diptera: Anthomyiidae). *Canadian Entomologist*, **88:** 634–639.

Wissinger, S.A. 1997. Cyclic colonization in predictably ephemeral habitats: A template for biological control in annual crop systems. *Biological Control*, **10:** 4–15.

Woin, N., Takow, J.A. and Kosga, P. 2005. Predatory ground-dwelling beetles

(Carabidae and Staphylinidae) in upland rice fields in North Cameroon. *International Journal of Tropical Insect Science*, **25(3):** 190–197.

Wolcott, G.N. 1937. An Animal Census of Two Pastures and a Meadow in Northern New York. *Ecological Monographs*, **7:** 1–90.

Wrase, D. 1992b. Revision der paläarktischen arten der Gattung Apotomus III. (Coleoptera, Carabidae, Apotomoni). *Deutsche Entomologische Zeitschrift* (N.F.), **39(1-3):** 69–95.

Wratten, S.D. 1987. The effectiveness of native natural enemies. *In:* A.J. Burn, T.H. Coaker and P.C. Jepson (eds). Integrated Pest Management. Academic London.

Yu, X., Liu, Y. and Axmacher, J.C. 2006. Field margin as rapidly evolving local diversity hotspots for ground beetles (Coleoptera: Carabiae) in northern China. *Coleopterists Bulletin*, **60:** 135–143.

Yücel, E. and Sahin, Y. 1988. Eskişehir ve Afyon Yöresi Bazi Carabidae (Coleoptera) Türlerinin Morfolojisi ve Ekolojisi Üzerine Çalişmalar I. *Anadolu Üniversitesi Fen Fakültes*, **2:** 25–37.

Zetto-Brandmayr, T. 1983. Life cycle, control of propagation rhythm and fecundity of *Ophonus rotundicollis* Fairm. et Lab (Coleoptera, Carabidae, Harpalini) as adaptation to the main feeding plant *Daucus carota* L. (Umbelliferae). *In:* P. Brandmayr, P.J. den Boer, E. Weber (eds). Ecology of Carabids: The Synthesis of Field Study and Laboratory Experiment. Centre Agric. Publ. Doc. Wageningen.

Zetto-Brandmayr, T. 1990. Spermophagous (seed eating) ground beetles: First comparison of the diet and ecology of the Harpaline genus *Harpalus* and *Ophionus* (Coleoptera: Carabidae). *In:* N.E. Stork (ed.). The Role of Ground Beetles in Ecological and Environmental Studies. Intercept Andover.

Zinner, K., Arnaut, D.A., Dognini, D.K. and Schmitte, K. 1991. Observations on the defensive reaction mechanism of *Pheropsophus aequinoctialis* (Carabidae). *Arquivos de Biologia e Tecnologia*, **34:** 185–193.

Index

C

M

N

U

V

W